Sports turf & amenity grasses

A MANUAL FOR USE AND IDENTIFICATION

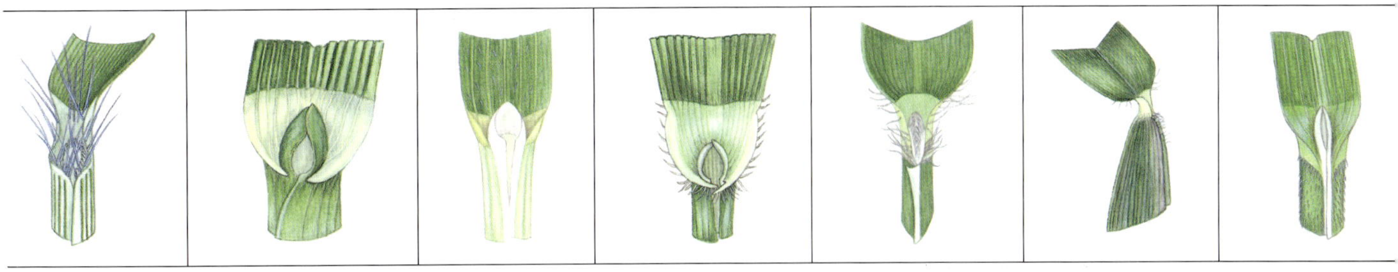

D E Aldous and I H Chivers

Colour illustrations by Enid Mayfield

LAND
LINKS

National Library of Australia Cataloguing-in-Publication entry

Aldous, David E., 1946– .
Sports turf and amenity grasses: a manual for use and identification.

Bibliography.
Includes index.
ISBN 0 643 06666 7.

1. Turfgrasses. 2. Turfgrasses – Handbooks, manuals, etc. 3. Turf management – Handbooks, manuals, etc. I. Chivers, Ian (Ian H.). II. Title.

635.9642

Available from:
Landlinks Press
150 Oxford Street (PO Box 1139)
Collingwood VIC 3066
Australia

Telephone: +61 3 9662 7666
Freecall: 1800 645 051 (Australia only)
Fax: +61 3 9662 7555
Email: publishing.sales@csiro.au
Web site: www.landlinks.csiro.au

Photographs by Ian H Chivers; colour illustrations by Enid Mayfield

Set in 9pt Frutiger Light
Cover and text design by James Kelly
Typeset by James Kelly
Printed in Australia by Brown Prior Anderson

Contents

About the authors iv

Preface v

Major sports turf and amenity grasses vi

Minor sports turf and amenity grasses vii

Further reading viii

- Helpful links viii

Turfgrass identification 1

- Turfgrass terminology 1
- Terms 2
- Using a visual turfgrass key 3

Key to common turfgrasses 4

MAJOR SPORTS TURF AND AMENITY GRASSES 25

Cool season grasses

- *Agrostis capillaris* Browntop bentgrass 26
- *Agrostis stolonifera* Creeping bentgrass 28
- *Festuca arundinacea* Tall fescue 30
- *Festuca longifolia* Hard fescue 32
- *Festuca rubra* ssp. *commutata* Chewing's fescue 34
- *Festuca rubra* ssp. *rubra* Red fescue 36
- *Lolium perenne* Perennial ryegrass 38
- *Poa annua* Annual bluegrass 40
- *Poa pratensis* Kentucky bluegrass 42
- *Poa trivialis* Rough stemmed bluegrass 44

Warm season grasses

- *Axonopus affinis* Narrow-leaf carpetgrass 46
- *Axonopus compressus* Broad-leaf Carpet grass 48
- *Buchloe dactyloides* Buffalo grass 50
- *Cynodon dactylon* Bermuda grass 52
- *Cynodon dactylon* × *C. transvaalensis* Hybrid bermudagrass 54
- *Dactyloctenium australe* Durban grass 56
- *Digitaria didactyla* Queensland blue couch 58
- *Eremochloa ophiuroides* Centipede grass 60
- *Paspalum dilatatum* Dallis grass 62
- *Paspalum notatum* Bahia grass 64
- *Paspalum vaginatum* Seashore paspalum 66
- *Pennisetum clandestinum* Kikuyu grass 68
- *Stenotaphrum secundatum* St Augustine grass 70
- *Zoysia japonica* Japanese lawngrass 72
- *Zoysia matrella* Manila grass 74
- *Zoysia tenuifolia* Korean lawngrass 76

MINOR SPORTS TURF AND AMENITY GRASSES 79

Cool season grasses

- *Agrostis alba* Redtop bentgrass 80
- *Agrostis canina* ssp. *canina* Velvet bentgrass 81
- *Austrodanthonia* spp. Wallaby grass 82
- *Bromus catharticus* Prairie grass 83
- *Cynosurus cristatus* Crested dog's tail grass 84
- *Dactylis glomerata* Orchard grass 85
- *Ehrharta calycina* Perennial veldt grass 86
- *Elytrigia repens* Twitch grass 87
- *Festuca ovina* Sheep fescue 88
- *Festuca pratensis* Meadow fescue 89
- *Holcus lanatus* Meadow softgrass 90
- *Lolium multiflorum* Italian ryegrass 91
- *Microlaena stipoides* Weeping grass 92
- *Phleum pratense* Timothy grass 93
- *Poa compressa* Canada bluegrass 94
- *Poa pratensis* × *P. arachnifera* Reveille bluegrass 95
- *Poa supina* Supina bluegrass 96
- *Puccinellia distans* Alkali grass 97

Warm season grasses

- *Bothriochloa macra* Red grass 98
- *Chloris gayana* Rhodes grass 99
- *Chloris truncata* Windmill grass 100
- *Chloris ventricosa* Tall chloris 101
- *Cynodon nlemfuensis* African stargrass 102
- *Cynodon transvaalensis* South African bermudagrass 103
- *Eragrostis cilianensis* Spreading lovegrass 104
- *Panicum laxum* Lax panic grass 105
- *Panicum repens* Torpedo grass 106
- *Paspalum distichum* Mercer grass 107
- *Paspalum nicorae* Brunswick grass 108
- *Sporobolus virginicus* Marine couch 109
- *Zoysia macrantha* Prickly couch 110

Index 111

About the authors

David E. Aldous is an Associate Professor in the School of Resource Management, Institute of Land and Food Resources, The University of Melbourne-Burnley Campus.

He is a Fellow of Parks and Leisure Australia, and is a member of the Australian Society of Horticultural Science, the International Society for Horticultural Science, and the International Turfgrass Society. In 1998 he was installed as the World President, of the International Federation of Park and Recreation Administration (IFPRA).

Ian H. Chivers is a private consultant and grass breeder/seed producer based in Melbourne, Victoria, Australia. He has worked in many parts of the world with both the cool season and warm season grasses. His particular expertise is in breeding and selection programmes for the less-well known grasses. This has led to the commercialisation of several species previously unknown for turf use. Within those species his work has led to the development of several varieties.

Ian is currently a Board member of the International Turfgrass Society, a contributor to several references and commonly seen as a conference speaker.

Dactyloctenium australe

Preface

To many people grass is unimportant. They walk on it, run on it, sometimes sleep on it, yet they pay little attention to it. This book will be of no interest to them. This book will be of great interest, however, to people who love to mow their lawn, repair the sports field, worry about the lie of the ball on the green or are interested in improving their environment by reducing pollution. This book is written to help identify the grasses that they mow, repair, fertilise and otherwise take care of. It aims to help identify the grasses in their turf and how they are likely to perform.

The turfgrass manager needs to be able to identify the grasses that are being used in order to tailor the management to its characteristics. The manager also needs to be able to select the best grass for a particular site to ensure the most suitable combination for the needs of that site. Turfgrass identification commonly depends heavily on the reproductive characteristics of the plant but often the flower head, or inflorescence, has been removed in the mowing process. In this book, we have used vegetative characters only, so that managers can identify the grass from a sample of the mown sward.

Answers to the question 'What are the most commonly found perennial grasses that are found in mown turf?' resulted in a list that could be divided into two groups: those that are sown or planted deliberately, and those that occur frequently as volunteer plants within a turf sward. We chose 26 species of grasses that in our view form the 'major' list. These grasses are widely sown or planted as the principal species as a turf or amenity grass cover. These species are described in some detail and are illustrated with botanically accurate illustrations to allow for very simple identification.

A further 31 species of grasses are included more briefly in our 'minor' list. These species are those that are not commonly sown or planted but which regularly appear as volunteer grasses in mown turf, and may comprise a significant percentage of the total turf cover. Some of these grasses have recently been dropped from mixtures where they had been used previously, such as Crested dog's tail grass (*Cynosurus cristatus*), whereas others are seen as emerging grasses which show potential to be used for sports turf or amenity purposes. Specifically excluded are the common annual weeds such as Barnyard grass (*Echinochloa crus-galli*), Summer grass (*Digitara sanguinalis*) or grass-like plants such as Toadrush (*Juncus bufonius*), as there are adequate means of identifying these species in other references.

An exception in the 'major' list is Annual bluegrass (*Poa annua*), which was included for several reasons: it is geographically widespread, it often forms a very high percentage of the turf cover, perennial forms are frequently found and some domestication and seed production has occurred with this species.

The handbook is divided into two parts, an identification method using a vegetative key; then an illustrated description of the 57 grasses we included in our 'major' and 'minor' lists. Because of the different common names used throughout the world, the authors have used the United States common names in their key and illustrated descriptions as these are the most widely known. For example *Poa annua* will be called Annual bluegrass, rather than Winter grass or Annual meadow grass.

Many people have contributed in the publication of this book. For the excellent illustrations thanks to Enid Mayfield, of the Geelong Botanic Gardens, City of Greater Geelong, Vic., Australia, who professionally captured the vegetative plant parts of the grasses. Acknowledgements also to Dr Don Loch, Department of Primary Industries, Cleveland, Qld, and Dr John Wilson, Tropical Agriculture, CSIRO Australia, St Lucia, Qld, for providing plants, slides and advice. We also acknowledge Bruce MacPhee, Peninsula Golf and Country Club, Frankston, Bruce Stephens, Nuturf Pty Ltd, John Donnelly, CSIRO Australia, St Lucia, Qld, David Vial, Integrated Open Space Services, Brisbane, Qld, Crystal Rose-Fricker, Turfseed Inc, USA and John Lloyd, Parliament House, Canberra, for grass specimens. Also special thanks to Ted Hamilton, CSIRO Publishing and Cyril Jerram, C.H. Jerram and Associates for their editing and production assistance.

David E Aldous
The Institute of Land and Food Resources
School of Resource Management,
The University of Melbourne, Burnley Campus
500 Yarra Boulevard, Richmond
Victoria 3121, Australia
email: daldous@unimelb.edu.au
Fax: (03) 9250 6885
Telephone: (03) 9250 6800

Ian H Chivers
PO Box 133, Sandringham
Victoria 3191, Australia
email: ian@nativeseeds.com.au
Fax: (03) 9598 7895
Telephone: (03) 9521 6473

Major sports turf and amenity grasses

Scientific name	Common name – Australia	Common name – USA	Common name – Europe	Common name – Asia/Africa
Cool season grasses				
Agrostis capillaris	Browntop bentgrass	Browntop bentgrass, Rhode Island bentgrass	Colonial bentgrass	Bentgrass
Agrostis stolonifera	Creeping bentgrass	Creeping bentgrass	Creeping bentgrass, Carpet bentgrass	Creeping bentgrass
Festuca arundinacea	Tall fescue	Tall fescue	Tall fescue	Tall fescue
Festuca longifolia	Hard fescue	Hard fescue	Hard fescue	Hard fescue
Festuca rubra ssp. *commutata*	Chewing's fescue	Chewing's fescue	Chewing's fescue	Chewing's fescue
Festuca rubra ssp. *rubra*	Creeping red fescue	Red fescue	Red fescue, Creeping fescue	Red fescue
Lolium perenne	Perennial ryegrass	Perennial ryegrass	Perennial ryegrass, English ryegrass, Ray grass	Perennial ryegrass
Poa annua	Winter grass	Annual bluegrass	Annual meadowgrass	Poa, Annual poa
Poa pratensis	Kentucky bluegrass	Kentucky bluegrass	Smooth stemmed/stalked meadowgrass, English meadowgrass, Fowl bluegrass	Bluegrass
Poa trivialis	Rough stemmed bluegrass	Rough stemmed bluegrass, Rough bluegrass	Rough stemmed/stalked meadowgrass	
Warm season grasses				
Axonopus affinis	Narrow-leaf carpetgrass, Matt grass	Common carpetgrass		Carpetgrass, Cow grass
Axonopus compressus	Broad-leaf carpetgrass, Tropical carpetgrass	Carpet grass		Cow grass
Buchloe dactyloides	US Buffalo grass	Buffalo grass		Buffalo grass
Cynodon dactylon	Couch grass, Green couch	Bermuda grass	Bermuda grass	Bermuda grass, Kweek grass, Indian doub
Cynodon dactylon × *C. transvaalensis*	Hybrid couch	Hybrid bermudagrass	Hybrid bermudagrass	Bermuda grass, Tif grass
Dactyloctenium australe	Sweet smother grass	Durban grass		
Digitaria didactyla	Queensland blue couch, Blue couch			Serangoon grass
Eremochloa ophiuroides	Centipede grass	Centipede grass, Chinese lawngrass		Chinese lawngrass
Paspalum dilatatum	Paspalum	Dallis grass, Water grass	Caterpillar grass	Dallis grass
Paspalum notatum	Bahia grass	Bahia grass		Bahia grass
Paspalum vaginatum	Saltwater couch	Seashore paspalum, Siltgrass, Sand knotgrass		Salt grass
Pennisetum clandestinum	Kikuyu grass	Kikuyu grass		Kikuyu grass, Kikuyo, Kikoejoe (South Africa)
Stenotaphrum secundatum	Buffalo grass	St Augustine grass		
Zoysia japonica	Zoysia	Japanese lawngrass		Zoysia
Zoysia matrella	Manila grass	Manila grass, Matrella zoysiagrass		Manila grass, Korean grass
Zoysia tenuifolia	Korean lawngrass	Korean lawngrass, Mascarene grass		Korean velvet grass

Minor sports turf and amenity grasses

Scientific name	Common name – Australia	Common name – USA	Common name – Europe	Common name – Asia/Africa
Cool season grasses				
Agrostis alba	Redtop bentgrass	Redtop bentgrass	Black bentgrass	
Agrostis canina ssp. *canina*	Velvet bentgrass	Velvet bentgrass	Brown bentgrass	
Austrodanthonia spp.	Wallaby grass			
Bromus catharticus	Prairie grass	Prairie grass	Rescue grass	Rescue grass (South Africa)
Cynosurus cristatus	Crested dog's tail	Crested dog's tail grass	Crested dog's tail grass, Dog's tail grass	
Dactylis glomerata	Cocksfoot	Orchard grass		
Ehrharta calycina	Perennial veldt grass	Perennial veldt grass	Perennial veldt grass	Common ehrharta (South Africa)
Elytrigia repens	Twitch grass	Twitch grass	Twitch grass, English couch, Scutch	
Festuca ovina	Sheep's fescue	Sheep fescue	Sheep fescue, Blue fescue	
Festuca pratensis	Meadow fescue	Meadow fescue	Meadow fescue, English bluegrass	
Holcus lanatus	Yorkshire fog	Meadow softgrass	Common velvetgrass	
Lolium multiflorum	Italian ryegrass	Italian ryegrass	Italian ryegrass, Common ryegrass, Annual ryegrass	
Microlaena stipoides	Weeping grass			
Phleum pratense	Timothy grass	Timothy grass	Timothy grass, Large leafed timothy grass	
Poa compressa	Canada bluegrass	Canada bluegrass	Flattened meadowgrass	
Poa pratensis × *P. arachnifera*	Reveille bluegrass	Reveille bluegrass		
Poa supina	Supina bluegrass	Supina bluegrass	Supina bluegrass, Prostrate meadow grass	
Puccinellia distans	Puccinellia	Alkali grass	Reflexed meadow grass, Reflexed poa	
Warm season grasses				
Bothriochloa macra	Red grass, Red-leg grass			
Chloris gayana	Rhodes grass	Rhodes grass		Hunyani grass, Koro-korosan (Asia)
Chloris truncata	Windmill grass, Star grass, Umbrella grass			
Chloris ventricosa	Tall chloris, Tall windmill grass			
Cynodon nlemfuensis	African stargrass			Star grass (South Africa)
Cynodon transvaalensis	South African couchgrass	Sth African bermudagrass, African bermudagrass, Royal blue grass		
Eragrostis cilianensis	Stink grass	Spreading lovegrass		Stink lovegrass (South Africa)
Panicum laxum	Panic grass, Lax panic grass			
Panicum repens	Torpedo grass, Couch panic	Torpedo grass	Creeping panic grass	Panic grass
Paspalum distichum	Water couchgrass	Mercer grass, Knotgrass		Knot grass, Water couch
Paspalum nicorae	Brunswick grass, Blue Dawn Brunswick grass	Brunswick grass		
Sporobolus virginicus	Marine couchgrass, Saltwater couch	Seashore dropseed, Beachdropseed		Saltwater smutgrass
Zoysia macrantha	Prickly couchgrass, Spiky couchgrass, Salt turf grass			

Further reading

Aldous, D.E. and P.S. Semos (1999). *Turfgrass identification and selection.* In International Turf Management Handbook. (Ed.) D.E. Aldous, Inkata Press, A Division of Butterworth-Heinemann, Port Melbourne, Chapter 2, pp. 19–47.

Beard, J.B. (1973) *Turfgrass: science and culture.* Prentice-Hall, Englewood Cliffs, N.J.

Burbridge, N.T. (1984) *Australian grasses*, Angus and Robertson, Sydney.

Gibbs, J. and Gibbs, R. *Grass (2001) Identification Manual—for everyone.* Native Grass Resources Group–South Australia. 103 pages.

Hafliger, E. and Scholz, H. (1980) *Grass weeds*, 1 and 2, Ciba-Geigy, Basle, Switzerland.

Hanson, A.A., Juska, F.V. and Burton, G.W. (1969) Species and varieties. In *Turfgrass science.* (Eds) A.A. Hanson and F.V. Juska (Agronomy, No. 14). American Society of Agronomy, Madison, Wisconsin, USA, pp. 370–409.

Harris, G.S. (1971) Identification of turf plants. In *Turf culture*, 2nd edn, pp.32–84, prepared by the New Zealand Institute for Turf Culture. Simon Printing Company, Palmerston North, New Zealand.

Hubbard, C.E. (1959) *Grasses*, 2nd edn. Penguin Books, Harmondsworth, Middlesex.

Hawley, P. (1989) *Grasses, Shire natural history*, C.I. Thomas & Sons (Haverfordwest) Ltd, Dyfed, UK.

Lambrechtsen, N.C. (1986) *What grass is that?* New Zealand Department of Scientific and Industrial Research. Government Printing Office, Wellington, New Zealand.

Lamp, C.A., Forbes, S.J. and Cade, J.W. (1990) *Grasses of temperate Australia*, Inkata Press, Melbourne.

Leinauer, B., Schulz, H., Bar, D. and Huber, A. (1997) *Poa supina*: a new species for turf. *International Turfgrass Society Research Journal*, 8, 345–351.

Mitchell, M. (1994) *Identification handbook for native grasses in Victoria.* Rutherglen Research Institute, Rutherglen, Vic.

Porter, M. (1999) *Texas A & M Blows Reveille for Hardy New Bluegrass.* April 2, page 30. http//agnews.tamu.edu/stories/SOIL/Apr0299.htm

Schulke, B. (2001) *Sown pasture grasses for the coastal Burnett.* Queensland Department of Primary Industries, June http:www.dpi.qld.gov.au/beef/3304.html

Skerman, P.J. and Riveros, F. (1990) *Tropical grasses*, (FAO Plant Production and Protection Series, No. 23), FAO, Rome.

Suttons Seeds (1962) *The identification of grasses by foliage*, Suttons Seeds Ltd, Reading, UK.

Tothil, J.C. and Hacker, J.B. (1983) *The grasses of Southern Queensland.* University of Queensland Press, St Lucia, Qld.

Van Wyk, E. and Van Oudtshoorn, F. (1999) *Guide to grasses of Southern Africa.* Briza Publications, Arcadia, South Africa.

Watson, J.R. (1994) Warm-season grasses. In *Natural turf for sport and amenity: Science and practice*, (Eds) W.A. Adams and R.J. Gibbs, CAB International, Wallingford, Oxon.

Watson, L. and Dallwitz, M.J. (1992) *The grass genera of the world.* CAB International, Wallingford, Oxon.

Wheeler, D.J.B., Jacobs, S.W.L. and Norton, B. (1990) *Grasses of New South Wales*, University of New England Monographs, Armidale, NSW.

Whittet, J.N. (1969) Turf grasses and their management. In *Pastures*, 2nd edn. Department of Agriculture, New South Wales, Sydney, pp. 605–616.

Helpful links

http://www.caf.wvu.edu/~forage/library/cangrass/

http://www.denislindsell.demon.co.uk/pasture/grasses/

http://www.uoguelph.ca/GTI/software/turfid/part4.htm

Turfgrass identification

Terminology

Before we can identify any grass, there is a need to understand several terms. These terms relate to the vegetative parts of a grass, rather than the flower head, as the head is not often available for use in turfgrass identification. The terms will be used in the visual key in the next section. Some of the more commonly used terms are given in the following diagram of a grass plant can be found in Figure 1.

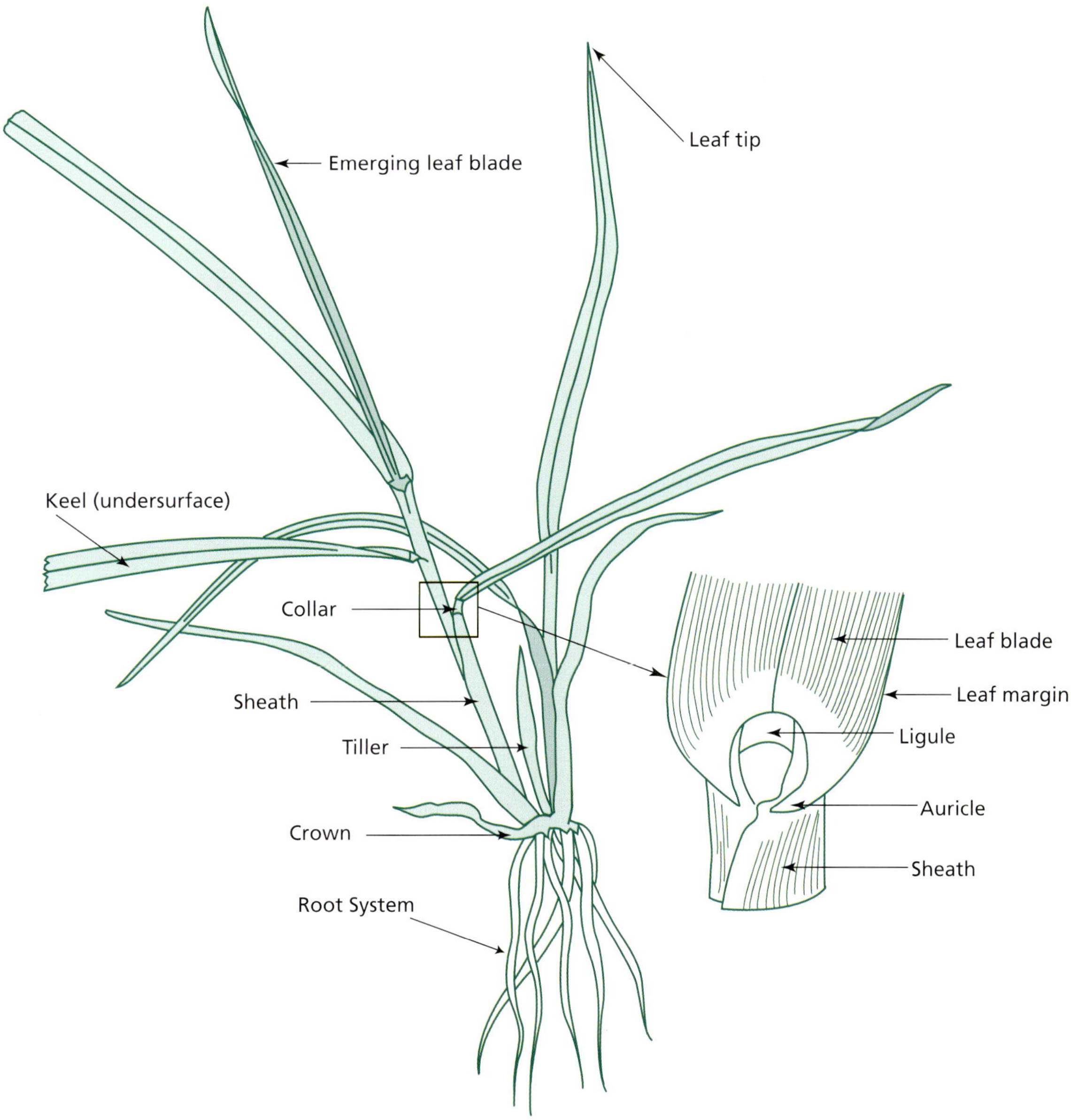

Figure 1. Structure of a typical turfgrass plant.

Terms

Acute
Sharp pointed (at the tip of the leaf blade).

Auricle
Refers to the outgrowths on the edge of the leaf that occur in some grasses, but may be absent on others. Auricles may also vary in size and shape. They occur at the junction of the leaf blade and sheath. They are usually either absent, rudimentary (small bud-like) or long and claw-like where they may overlap.

Blade
Extended upper portion of the leaf beyond the sheath. Leaf tip, leaf surface, margins and colour often can assist in identification. Leaf tips are often presented as being acute, blunt or rounded, boat or canoe shaped or tapering to a point.

Bunch type
Erect growth of young vegetative stems to form a crown.

Cilia
A hair or bristle

Ciliate
Fringed with fine, soft hairs, often on the margins of the leaf.

Collar
Strengthening tissue at the back of the ligule, located immediately between the blade and above the leaf sheath. Among the grasses, collars may be broad, or narrow, continuous or divided, oblique, as well as being completely covered with hairs, not hairy or only partially hairy at the outer margins of the collar.

Glabrous
Not hairy or pubescent.

Growth habit
Expression of how the turfgrass grows and develops. Growth habits are often represented as a bunch habit of growth (tillers emerging from the crown), stoloniferous (horizontal stem, usually prostrate and above ground level, commonly rooting at the nodes), rhizomatous (horizontal, non-green, underground creeping stem) and bulb-like (swollen base).

Keel
Of some leaf shapes on the lower surface of the leaf where the mid-vein protrudes like the keel of a boat or canoe.

Leaf margin
Refers to the margin, or edge of the leaf, and although variable within certain limits in any species, can be of considerable value in separating certain grasses. Leaf margins are often represented as either scabrous.

Leaf shape
A cross sectional representation of a turfgrass leaf can often assist in identification. Leaf shapes are often classed as being flat, v-shaped, bristle-like or rolled inward. Sometimes leaf shape can refer to characters of the upper and lower surfaces of the leaf. Commonly they are keeled with a projecting ridge on the undersurface the leaves, or ribbed with ridges of various shapes running on the upper surface of the leaf.

Leaf surface
Refers to leaf topography and can also be useful in grass identification. Leaf surfaces are often represented as being pubescent or glabrous.

Ligule
Small membrane or ring of hairs that occurs at the inner junction of the leaf blade and sheath. There are three basic ligule types, they are either absent, membranous or are a fringe of hairs.

Ligule shape
Ligule shape can also be useful in identification. Ligule shapes have been classified as truncate (cut off square at the summit of the ligule), acute (sharp pointed), round (semi-circular) or acuminate (tapering to a point).

Ligule margin
The margin of the membranous ligule can also assist in identification. Ligule margins may be entire (smooth), distinctly notched (on either side of the ligule) or toothed (ragged at summit of ligule).

Margin
Refers to the outside edge of the leaf blade.

Membranous
A description of a ligule being of one continuous enveloping structure, not divided into individual hairs. Usually thin layered and pliable, almost transparent and colourless.

Midrib
Central vein of the blade of the leaf, often forming a pronounced ridge on the upper surface and a keel below.

Node
Point on the stem where leaves are inserted.

Perennial
Plant that lives for an indefinite number of growing seasons.

Petioled
A petiole may occur at the junction of the blade and the sheath. It is usually not defined but may be for some grasses where a distinct joining section occurs to connect the sheath and the leaf blade. In this case the leaf is said to be petioled.

Pubescent
Very short hairs (up to 0.5 mm).

Rhizome
Refers to a horizontal, usually non-green, underground creeping stem. These frequently give rise to new or 'daughter' plants at some distance from the 'mother' plant.

Ribs
Ridges of various shapes running on the upper surface of the leaf.

Scabrous
Rough and harsh to the touch; often show short stiff bristles or saw-like teeth on the part concerned.

Serrate
Saw-like teeth.

Sheath
Refers to that part of the leaf, originating from the node that surrounds the stem. Some sheaths, when observed in cross section, appear to be have been compressed and this character (and the degree of compaction) are used in identification. Some sheaths are split vertically and are overlapping, others are not split at all, whereas others are partially split. Hairiness (pubescence), or the lack of hair (glabrousness), and the roughness of the surface and margins can also be helpful in identification. The colour of the sheath, particularly at the base, is also regularly used as an aid in identification.

Spreading
Growth habit exhibited by development of both/either stolons and rhizomes

Stolon
Refers to the horizontal stem, usually prostrate or trailing at ground level, always above the ground, often rooting at the nodes.

Tiller
Lateral shoot that arises within the sheath from the stem.

Vernation
Vernation can be determined by taking a thin cross section of the tiller (taken at the 2–3 leaf stage from the top of the tiller) with a hand scalpel, and examining this section under a 10 × hand lens or stereomicroscope. Vernation can be either folded or rolled in the bud.

Using a visual turfgrass key

A visual key is an artificial guide used to separate those grasses in the key. Hold a specimen plant with you as you work through the choices posed by the key. Materials that you will need are a 10 × hand lens, or a stereomicroscope, a clean new razor blade or scalpel, and some tweezers to separate out the plant parts. The first decision is to determine the vernation of the grass (i.e. is the grass is either rolled or folded in the bud). Once you have decided this, look at the presence or absence of auricles. Now you are in a position to enter the key. Follow the prompts and move through the key.

The choices are illustrated to enable easy separation between the various plant parts. Once you have made a choice, follow the arrow to the next section, where you may be asked to make a further choice. Continue this process until the features end with the name of a grass. If the features do not correspondence to your specimen, go back into the key to the point where you are most confident and look at the alternative or alternatives. It is often best to key out a number of grass plants of the same species to clarify your decision.

When using the key, remember that there are many environmental and cultural factors that can affect the characteristics of the grass plant, and certain grasses, because of their size, may be not easily identified.

This key describes only those grasses included in this book. While all care has been taken to encourage a positive identification, the floral characters can also be used when they are expressed.

Key to common turfgrasses

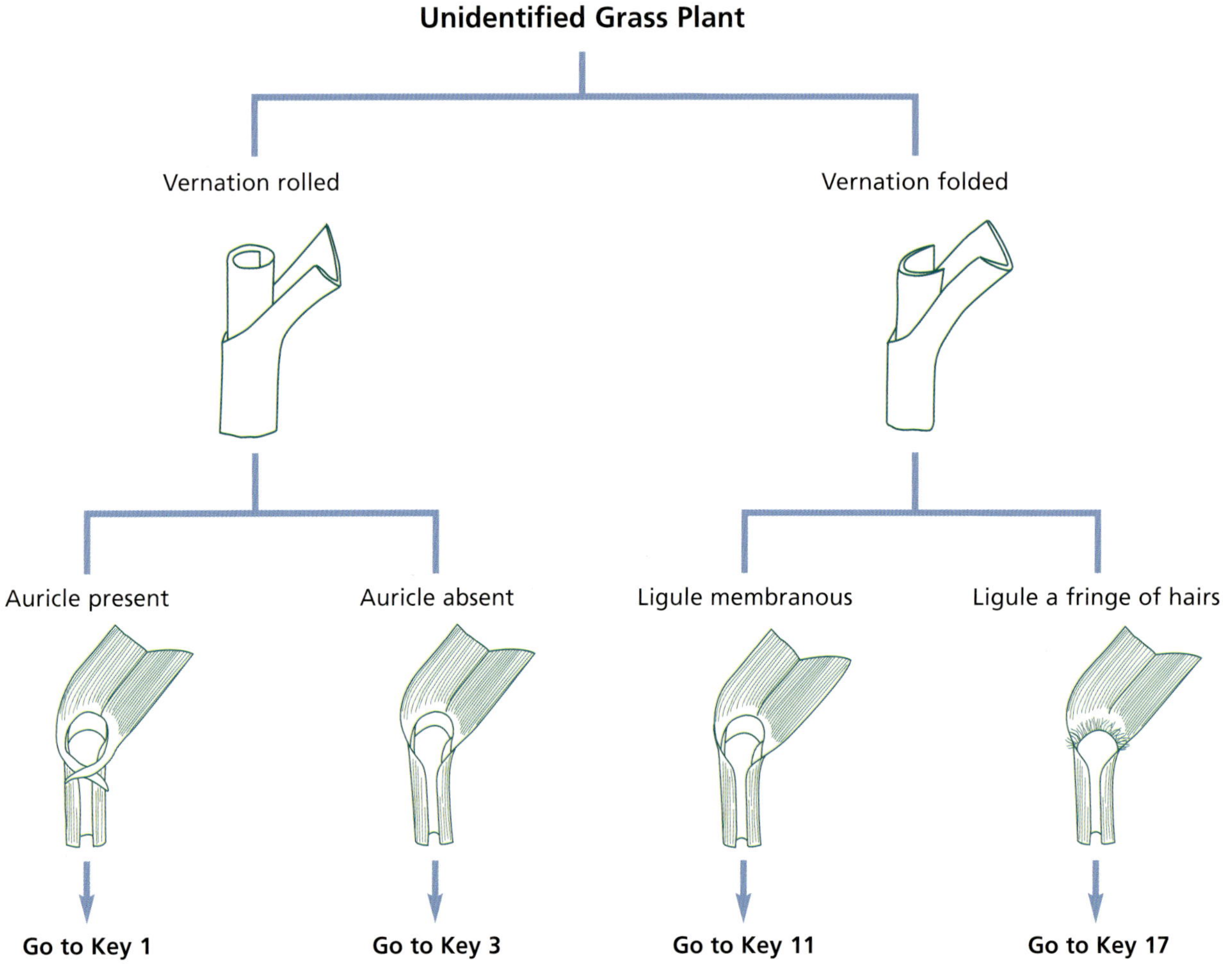

Key 1

Vernation rolled, auricle present

Collar divided

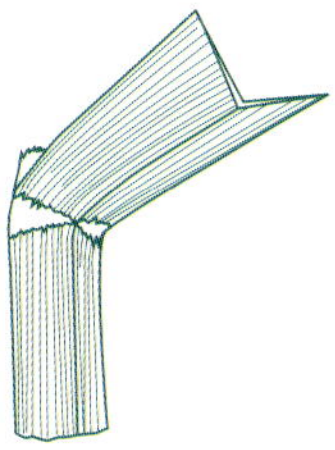

Growth habit – bunch type

Tall fescue
(*Festuca arundinacea*)
See page 30

Growth habit – rhizomatous

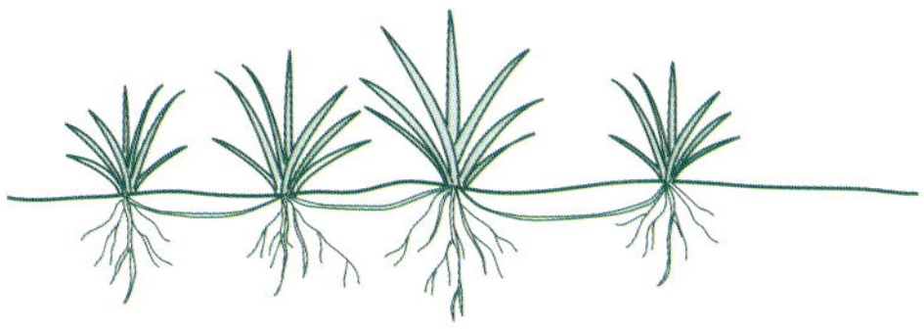

Twitch grass
(*Elytrigia repens*)
See page 87

Collar continuous

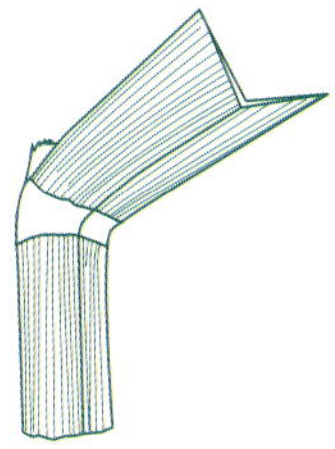

Go to Key 2

Key 2

Vernation rolled, auricle present, collar continuous

Leaf blade hairless

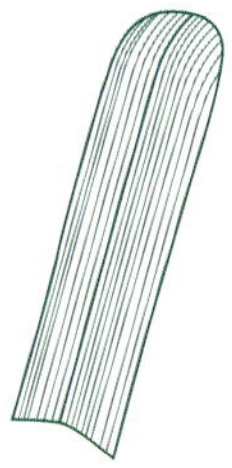

Perennial veldt grass
(*Ehrharta calycina)*
See page 86

Leaf blade with glossy lower surface

Leaf margins scabrous

Meadow fescue
(*Festuca pratensis*)
See page 89

Leaf margins smooth

Italian ryegrass
(*Lolium multiflorum)*
See page 91

Key 3

Vernation rolled, auricle absent

Collar divided

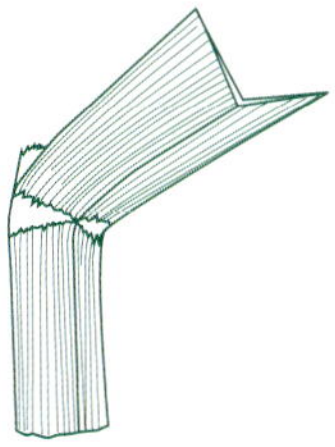

Ligule membranous

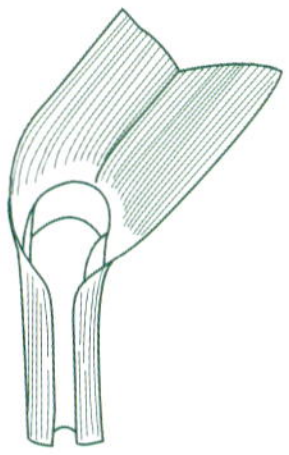

Leaf upper surface hairless and ridged

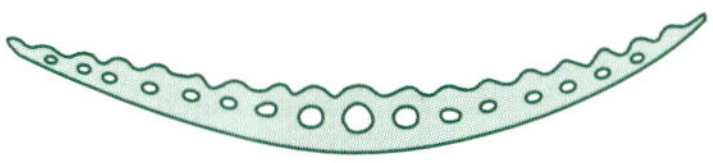

Browntop bentgrass
(*Agrostis capillaris*)
See page 26

Leaf upper surface sparsely hairy and smooth

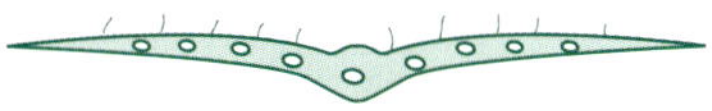

Queensland blue couch
(*Digitaria didactyla*)
See page 58

Ligule a fringe of hairs

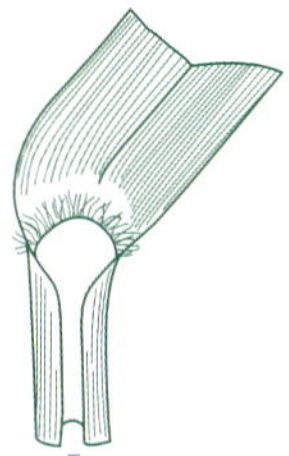

African stargrass
(*Cynodon nlemfuensis*)
See page 102

Collar continuous

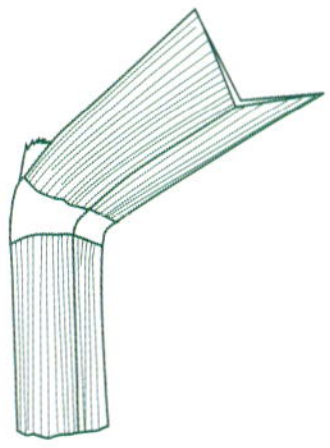

Go to Key 4

Key 4

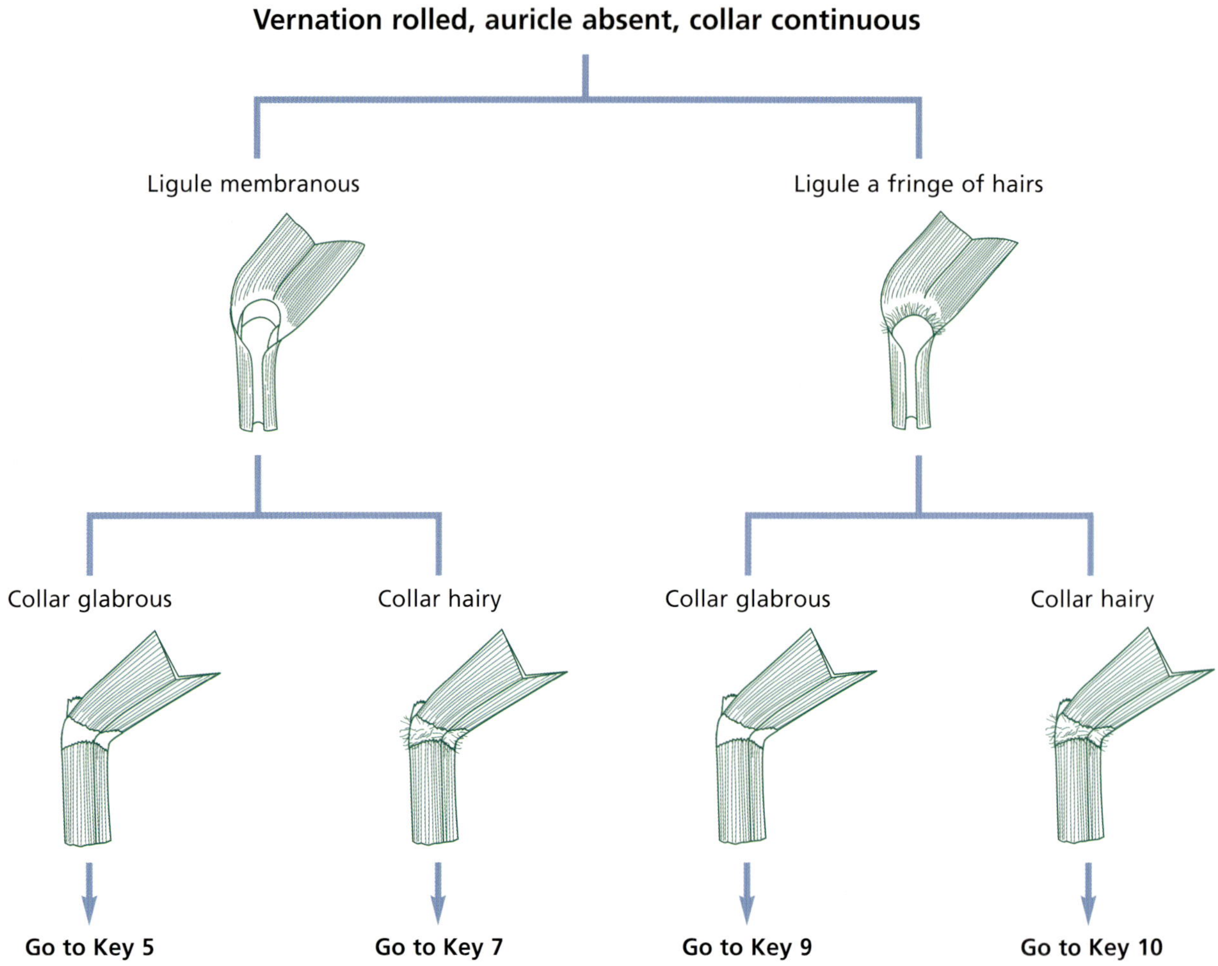
Vernation rolled, auricle absent, collar continuous
Ligule membranous
Ligule a fringe of hairs
Collar glabrous
Collar hairy
Collar glabrous
Collar hairy
Go to Key 5
Go to Key 7
Go to Key 9
Go to Key 10

Key 5

Vernation rolled, auricle absent, collar continuous, ligule membranous, collar glabrous

Bunch habit of growth

"Spreading" habit of growth

Go to Key 6

Sheath glabrous

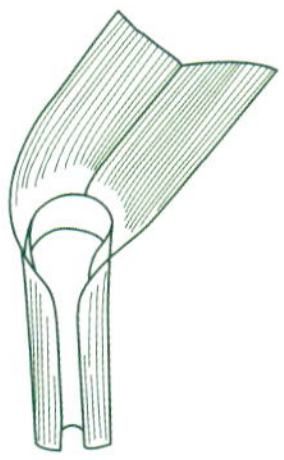

Leaf upper surface hairless and ridged

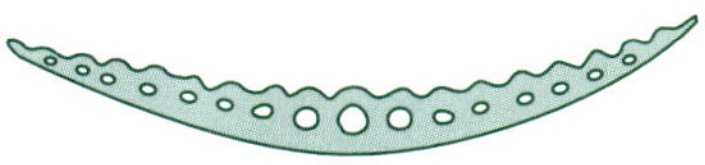

Redtop bentgrass
(*Agrostis alba*)
See page 80

Leaf under surface sparsely hairy and with distinct keel

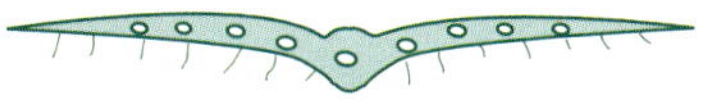

Timothy grass
(*Phleum pratense*)
See page 93

Sheath hairy

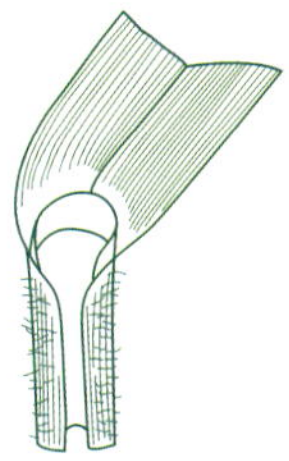

Prairie grass
(*Bromus catharticus*)
See page 83

Key 6

Vernation rolled, auricle absent, collar continuous, ligule membranous, collar glabrous, spreading habit

Leaf blades narrow < 3 mm

Leaf blades wide > 6 mm

Bahia grass
(*Paspalum notatum*)
See page 64

Collar broad

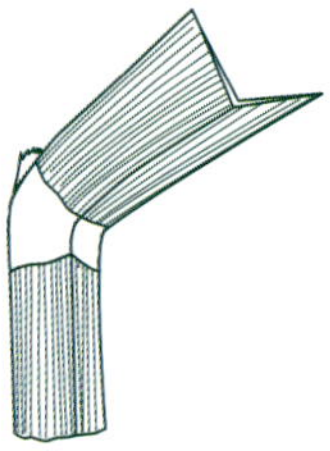

Collar narrow

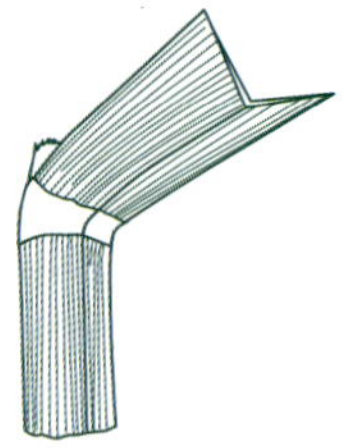

Ligule margins toothed

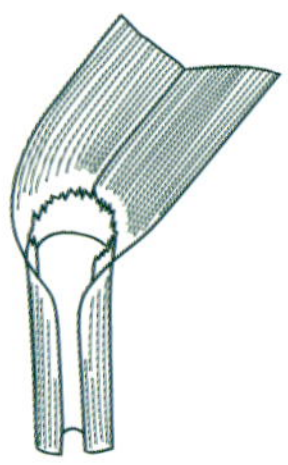

Ligule margins smooth

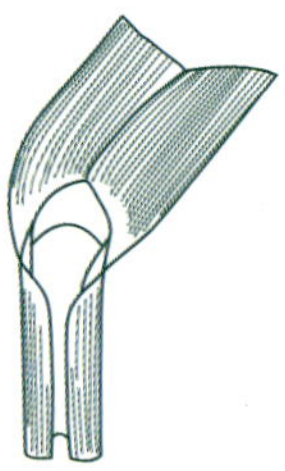

Ligule short

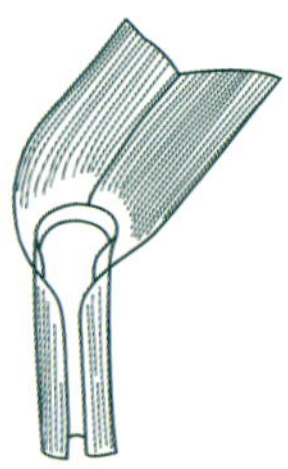

Ligule long

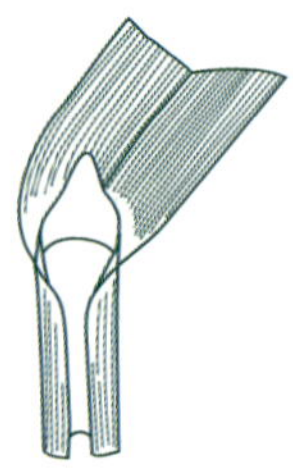

Creeping bentgrass
(*Agrostis stolonifera*)
See page 28

Velvet bentgrass
(*Agrostis canina* ssp. *canina*)
See page 81

Browntop bentgrass
(*Agrostis capillaris*)
See page 26

Redtop bentgrass
(*Agrostis alba*)
See page 80

Key 7

Vernation rolled, auricle absent, collar continuous, ligule membranous, collar hairy

Collar broad

- Sheath hairy at margins

  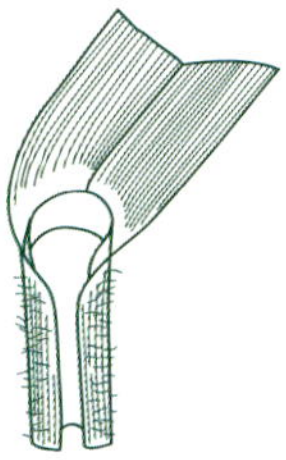

  Torpedo grass
  (*Panicum repens*)
  See page 106

- Sheath glabrous

  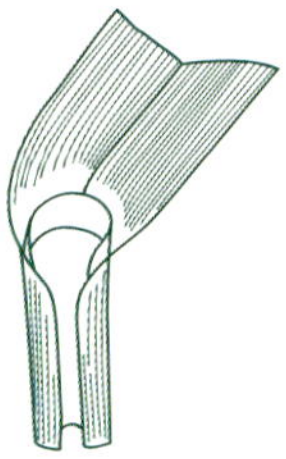

  - Leaf blades upper surface hairy

    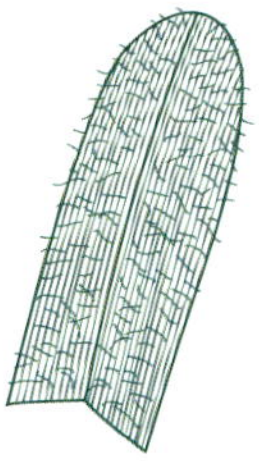

    Redgrass
    (*Bothriochloa macra*)
    See page 98

  - Leaf blades upper surface glabrous

    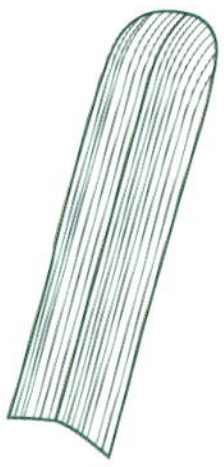

    Dallis grass
    (*Paspalum dilatatum*)
    See page 62

Collar narrow

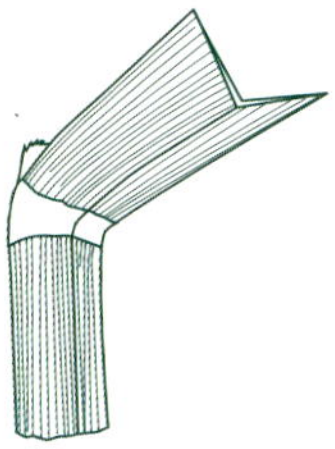

Go to Key 8

Collar indistinct

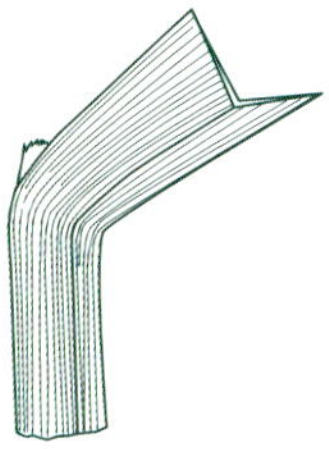

Marine couch
(*Sporobolus virginicus*)
See page 109

Key 8

Vernation rolled, auricle absent, collar continuous, ligule membranous, collar hairy, collar narrow

Leaf upper surface hairy

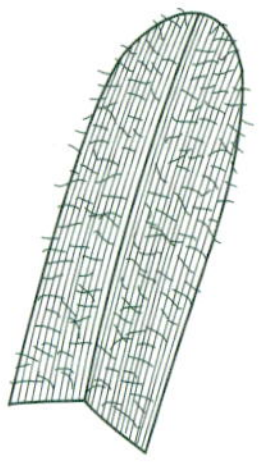

Ligule long

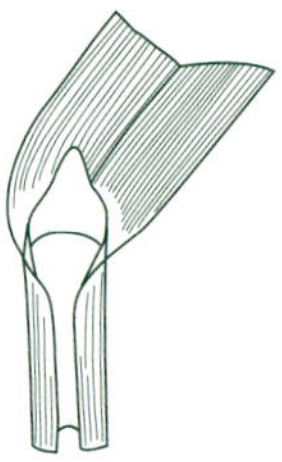

Meadow softgrass
(*Holcus lanatus*)
See page 90

Ligule short

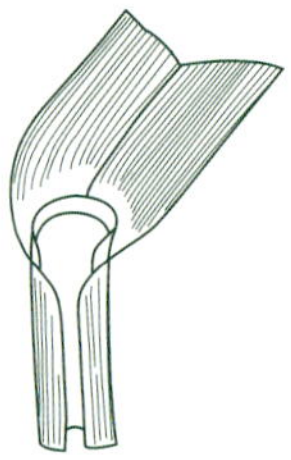

Ligule fringed by hair

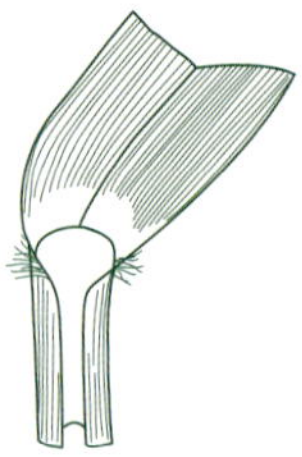

Weeping grass
(*Microlaena stipoides*)
See page 92

Ligule margins smooth

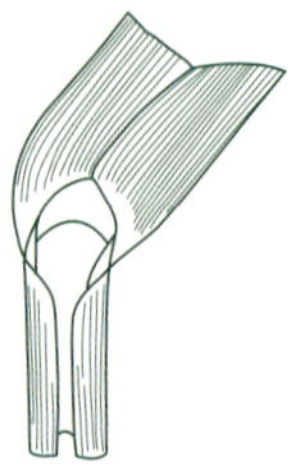

Seashore paspalum
(*Paspalum vaginatum*)
See page 66

Leaf upper surface glabrous

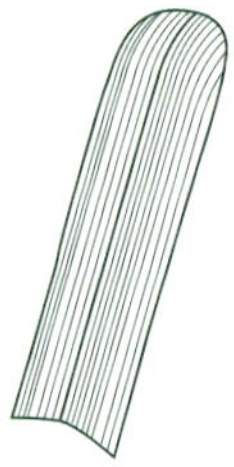

Mercer grass
(*Paspalum distichum*)
See page 107

Key 9

Vernation rolled, auricle absent, collar continuous, ligule a fringe of hairs, collar glabrous

- Leaf blades wide > 5–8 mm
  - Brunswick grass (*Paspalum nicorae*) **See page 108**
- Leaf blades narrow < 5–8 mm
  - Leaf blades not hairy at margins

    - Fine to medium leaf texture, usually around 2 to 3 mm width
      - Manila grass (*Zoysia matrella)* **See page 74**
    - Very fine, narrow leaves, usually less than 1.5 mm width
      - Korean lawngrass (*Zoysia tenuifolia*) **See page 76**
  - Leaf blades hairy at margins

    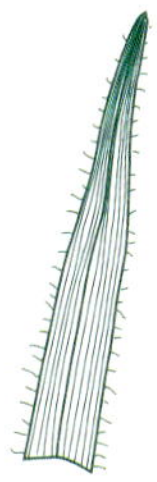

    - Buffalo grass (*Buchloe dactyloides*) **See page 50**

Key 10

Vernation rolled, auricle absent, collar continuous, ligule a fringe of hairs, collar hairy

Collar narrow

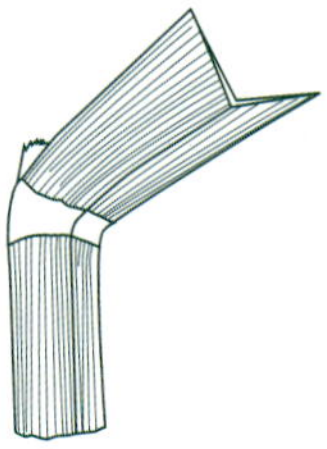

Collar broad

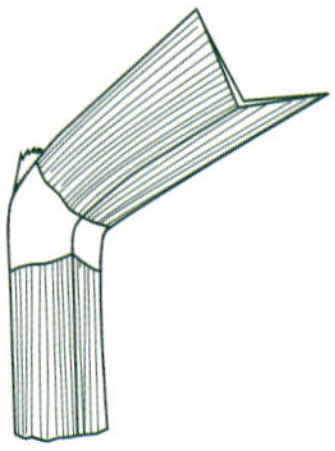

Blades stiff and durable

Japanese lawngrass
(*Zoysia japonica*)
See page 72

Blades soft and pliable

Lax panic grass
(*Panicum laxum*)
See page 105

Spreading growth habit

Prickly couch
(*Zoysia macrantha*)
See page 110

Bunch growth habit

Spreading lovegrass
(*Eragrostis cilianensis*)
See page 104

Key 11

Vernation folded, ligule membranous

Bunched growth habit

- Collar divided

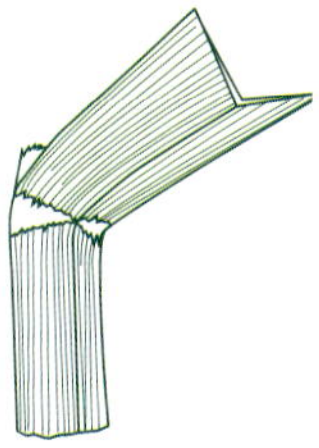

Go to Key 12

- Collar continuous

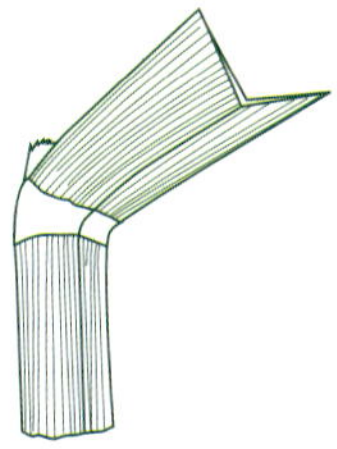

Go to Key 13

"Spreading" growth habit

Go to Key 15

Key 12

Vernation folded, ligule membranous, bunched growth habit, collar divided

Collar narrow

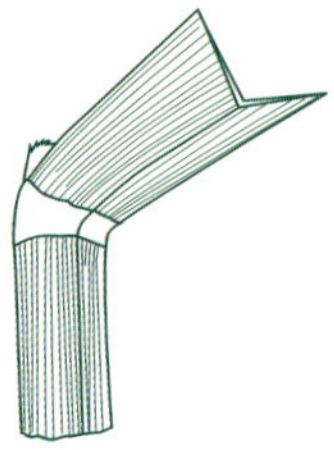

Ligule glabrous

Hard fescue
(*Festuca longifolia*)
See page 32

Ligule minutely hairy

Sheep fescue
(*Festuca ovina*)
See page 88

Collar broad

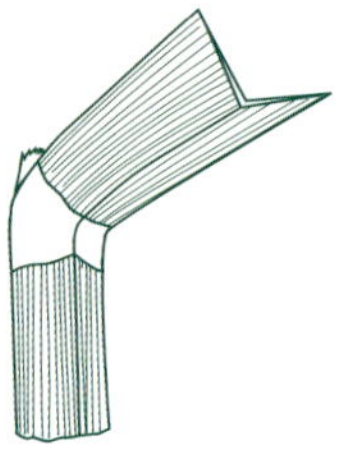

Ligule short, serrated

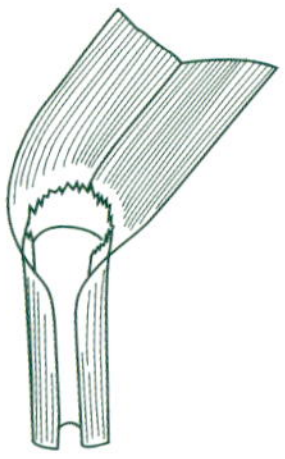

Windmill grass
(*Chloris truncata*)
See page 100

Ligule long, acute

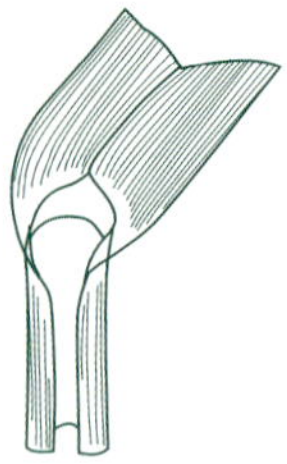

Orchard grass
(*Dactylis glomerata*)
See page 85

Key 13

Vernation folded, ligule membranous, bunched growth habit, collar continuous

Collar broad

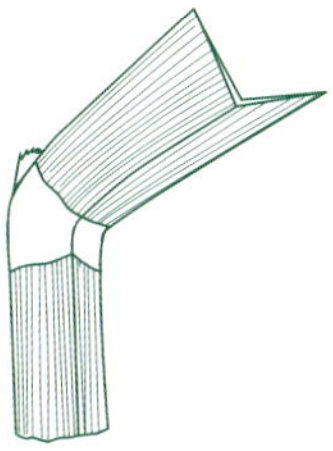

Ligule short, serrated

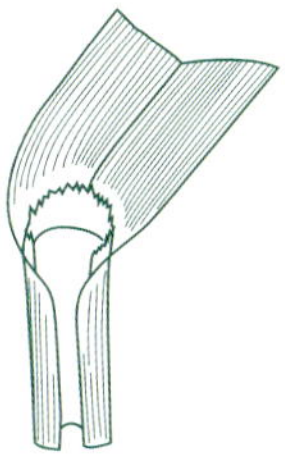

Tall chloris
(*Chloris ventricosa*)
See page 101

Ligule long, entire

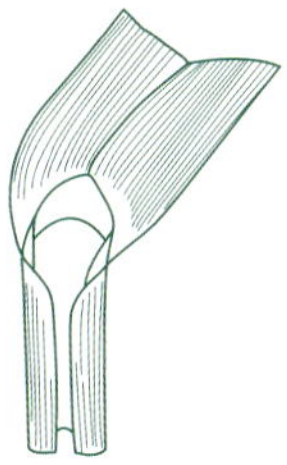

Annual bluegrass
(*Poa annua*)
See page 40

Collar narrow

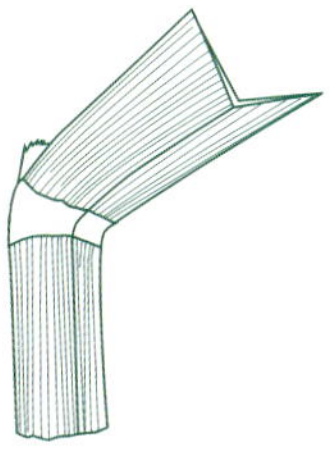

Leaf blade bristle-like

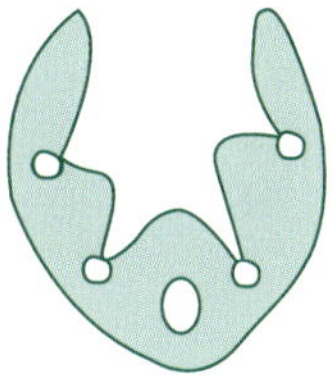

Chewings fescue
(*Festuca rubra* spp. *commutata*)
See page 34

Leaf blade flat

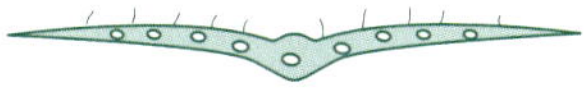

Go to Key 14

Key 14

Vernation folded, ligule membranous, bunched growth habit, collar continuous, collar narrow, leaf blade flat

Ligule not toothed (entire)

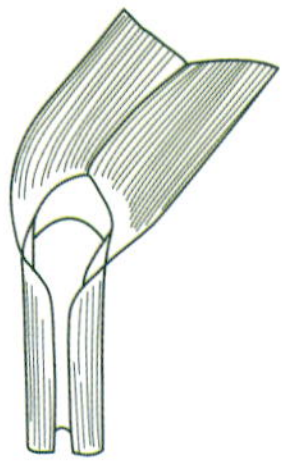

Leaf blade with distinct veins

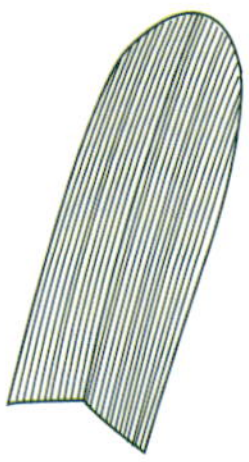

Crested dog's tail grass
(*Cynosurus cristatus*)
See page 84

Leaf blade without veins

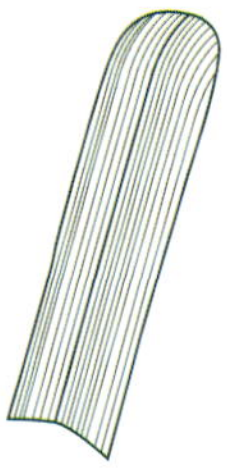

Alkali grass
(*Puccinellia distans*)
See page 97

Ligule toothed near apex

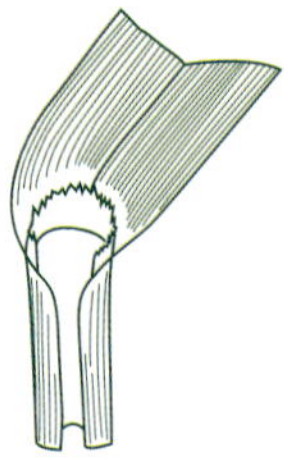

Perennial ryegrass
(*Lolium perenne*)
See page 38

Key 15

Vernation folded, ligule membranous, spreading growth habit

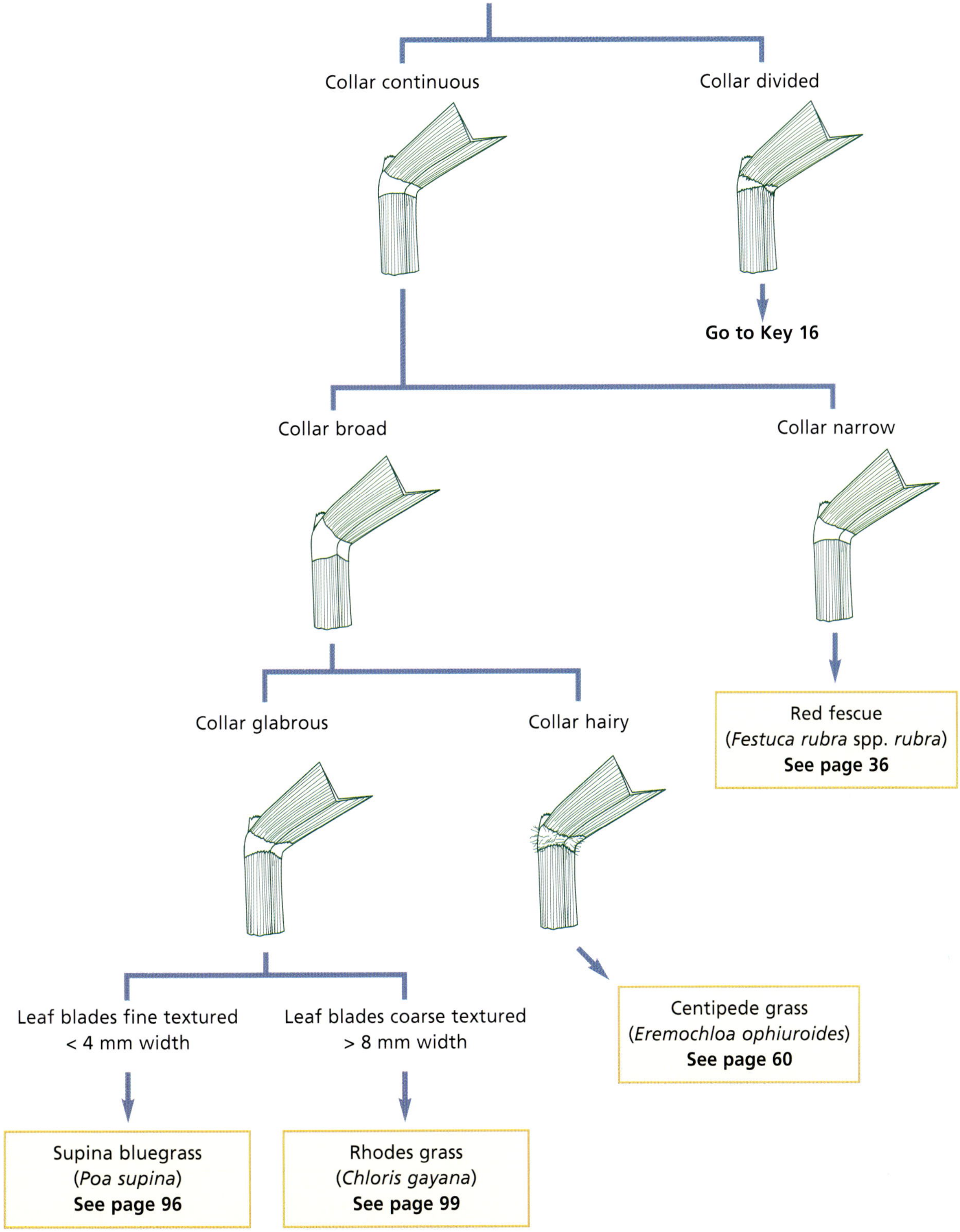

Key 16

Vernation folded, ligule membranous, spreading growth habit, collar divided

Ligule long

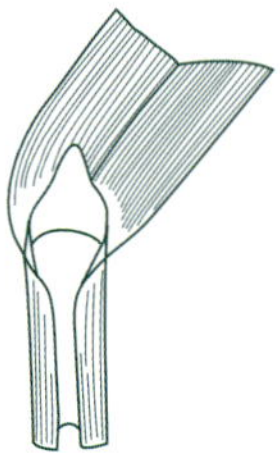

Rough stemmed bluegrass
(*Poa trivialis*)
See page 44

Ligule short

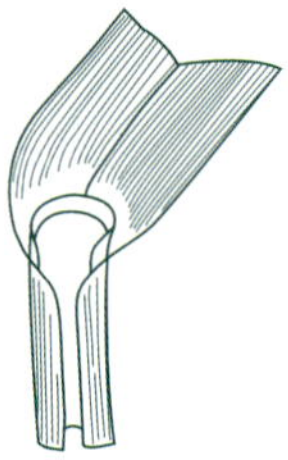

Collar broad

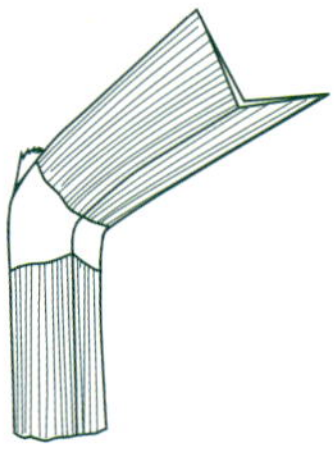

Leaf undersurface shiny

Kentucky bluegrass
(*Poa pratensis*)
See page 42

Leaf undersurface dull

Reveille bluegrass
(*Poa* hybrid)
See page 95

Collar narrow

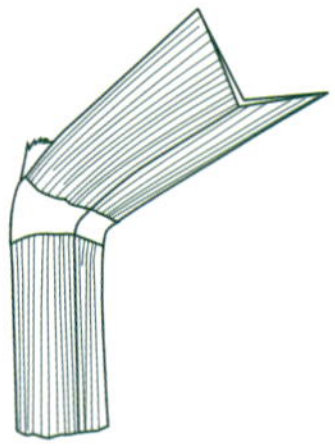

Canada bluegrass
(*Poa compressa*)
See page 94

Key 17

Vernation folded, ligule a fringe of hairs

Collar glabrous

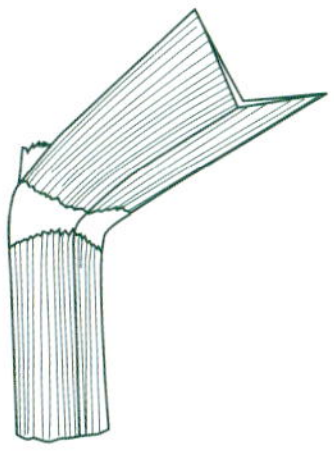

Collar hairy

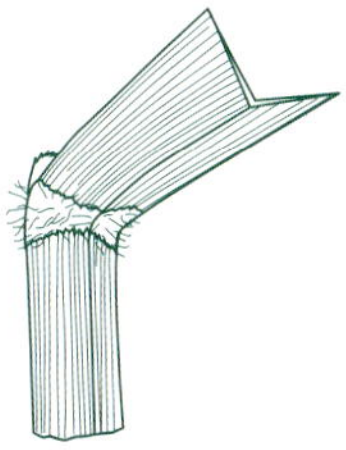

Go to Key 19

Sheath very hairy

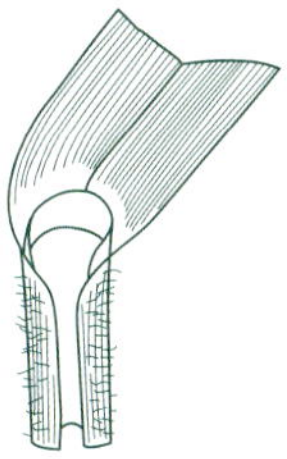

Sheath not hairy or only sparsely hairy

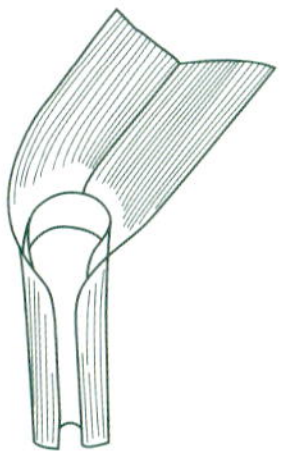

Go to Key 18

Leaf blades rounded, often undulating at the margins and glandular hairs towards base of leaf

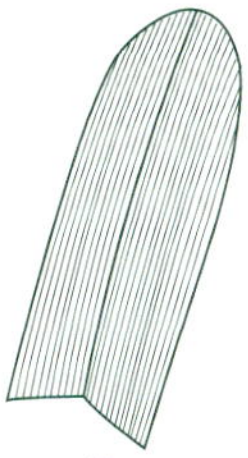

Broad leaf Carpet grass
(*Axonopus compressus*)
See page 48

Leaf blades tapering, usually straight

Narrow leaf carpetgrass
(*Axonopus affinis*)
See page 46

Key 18

Vernation folded, ligule a fringe of hairs, collar glabrous, sheath not hairy or only sparsely hairy

- Leaf blades not hairy at margins
  - Hairy at base of leaf

    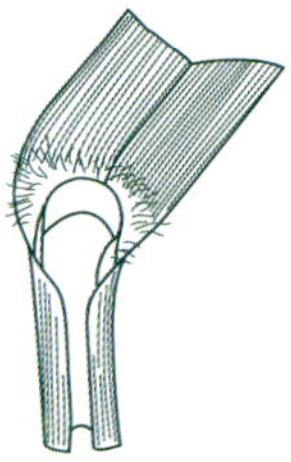

    - Collar broad and obvious

      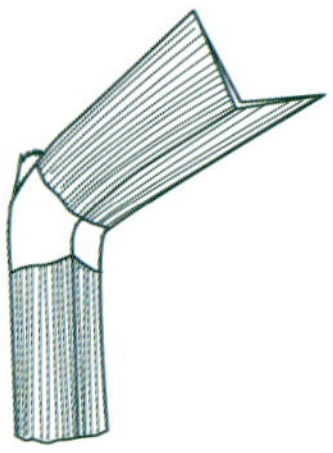

      St Augustine grass
      (*Stenotaphrum secuadatum*)
      See page 70

    - Collar narrow

      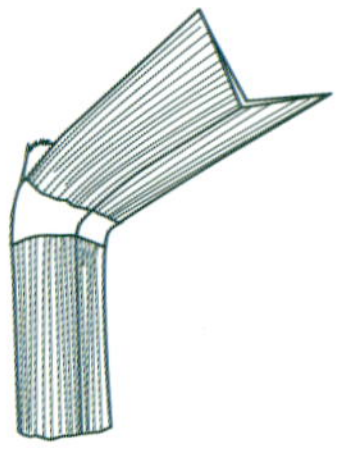

      Bermuda grass
      (*Cynodon dactylon*)
      See page 52

  - Not hairy at base of leaf

    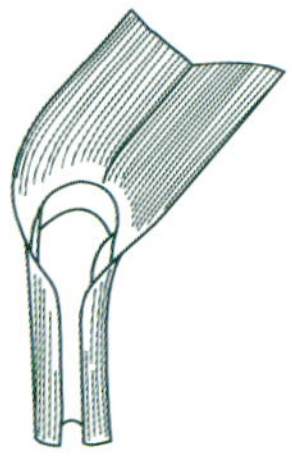

    Hybrid bermudagrass
    (*Cynodon dactylon* × *C. transvaalensis*)
    See page 54

- Leaf blades hairy at the margins

  Durban grass
  (*Dactyloctenium australe*)
  See page 56

Key 19

Vernation folded, ligule a fringe of hairs, collar hairy

"Spreading" growth habit
Rhizome / Stolon

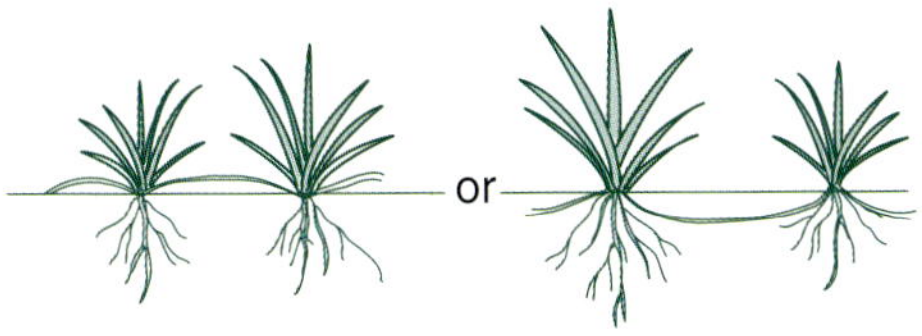

- Leaf blades broad, hairy

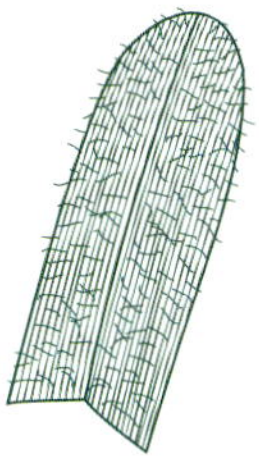

Kikuyu grass
(*Pennisetum clandestinum*)
See page 68

- Leaf blades narrow, glabrous

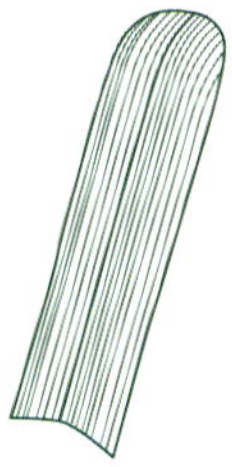

South African bermudagrass
(*Cynodon transvaalensis*)
See page 103

Bunched growth habit

Wallaby grass
(*Austrodanthonia* spp.)
See page 82

Major

SPORTS TURF AND AMENITY GRASSES

The following grasses are widely sown or planted deliberately as the principal species in a turf or as an amenity grass cover. These species are described in detail and are illustrated with botanically accurate illustrations to allow for simple identification.

BOTANICAL NAME *Agrostis capillaris*

COMMON NAME Browntop bentgrass

OTHER NAMES Colonial bentgrass

DESCRIPTION Upright perennial, fine textured turf of high density; Colour usually mid-green to grey-green; Little creeping tendency, but can spread by rhizomes. Sometimes called *Agrostis tenuis.*

COMMON USES Fairway oversowing, golf surrounds, greens, fine lawns, shaded lawns.

OTHER USES Golf tees, golf aprons, bowls/croquet, erosion control.

POSITIVES Prefers poor acid type soils; Very fine leaves and upright nature mean that this grass can be used for fine turf surfaces such as golf greens; Is difficult to divot when established; Can tolerate very low mowing.

NEGATIVES Drought and heat tolerance poorer than for many other cool season species, but will regenerate in the following season in most cases; Thatching tendency if allowed to grow too tall; Only moderate wear tolerance; Some disease problems.

COMMENTS Widely used in seed mixtures since it can fill gaps and improve the density of a turf surface; Can become dominant if close mowing is continuous; Mixes well with Red fescue to form fine turf surface. Inferior to Creeping bentgrass and Velvet bentgrass for greens.

IDENTIFYING CHARACTERS Auricles – absent; Blades – tapering, flat, scabrous margin, upper surface glabrous and distinctly ridge, slightly keeled on the lower surface; Collar – narrow, divided, glabrous; Ligule – membranous, 0.3–1.2 mm long, truncate, entire; Sheath – round, hairless, dark green; Vernation – rolled.

SOWING, GROWING AND MOWING

	New sowing rate		Oversowing rate		Stolonising rate		Mowing Heights	
	g/m²	lbs/1000 sq ft	g/m²	lbs/1000 sq ft	kg/100 m²	lbs/1000 sq ft	mm	inches
High	12	2.5	10	2.0	40	82	30	1.2
Low	3	.6	2	.4	20	41	2	.08

TOLERANCE

High temperature

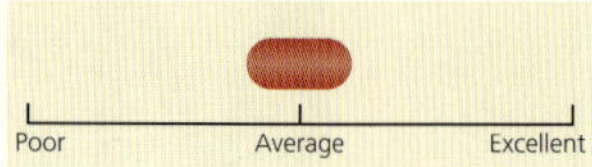

Drought

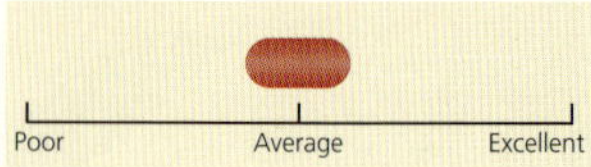

Close mowing

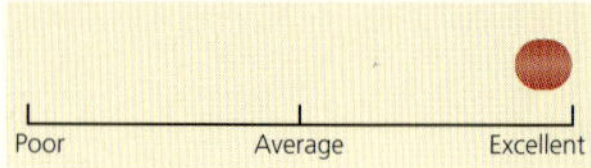

Low fertility

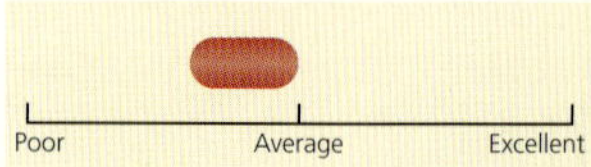

Frost

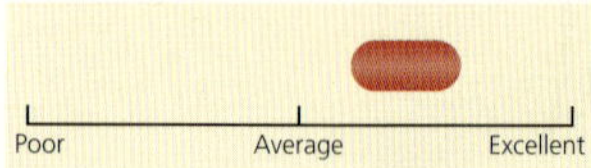

Shade

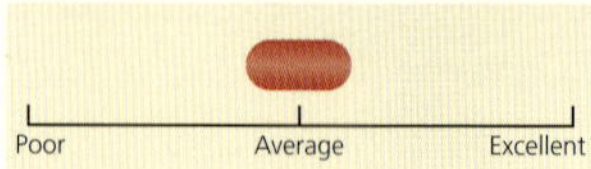

Wet soil

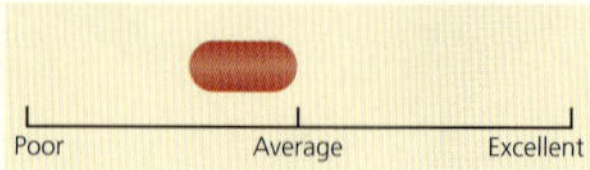

Wear

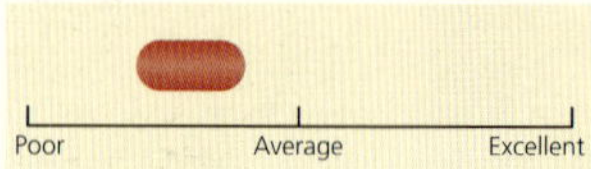

Salinity

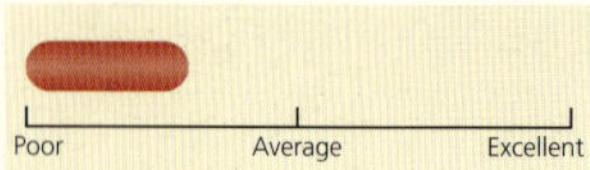

×1.0

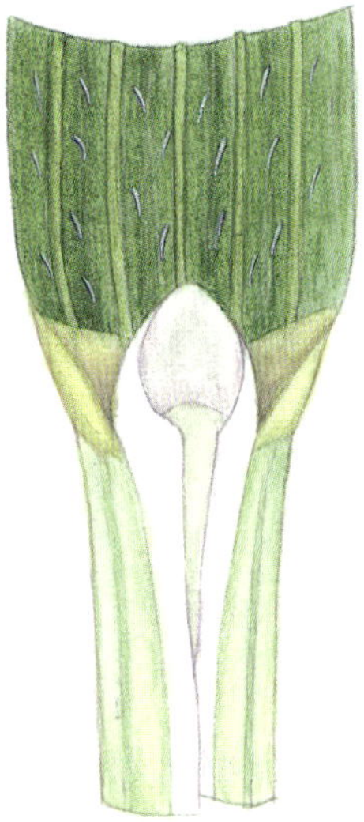

×20.0

BOTANICAL NAME *Agrostis stolonifera*

COMMON NAME Creeping bentgrass

OTHER NAMES Carpet bentgrass

DESCRIPTION High density turfgrass with narrow leaves; Colour usually green to deep-green; Many cultivars have long, creeping stolons; Stolons are usually soft and leafy; Rhizomes are present occasionally; Leaves are not shiny. Sometimes known as *Agrostis palustris*.

COMMON USES Greens, golf tees, golf aprons, fairway oversowing, golf surrounds.

OTHER USES Greens oversowing, fine lawns, bowls/croquet.

POSITIVES Once established it forms a uniform surface of high quality; Can tolerate close mowing; Will regenerate from rhizomes and stolons after damage; Can root from nodes.

NEGATIVES Requires a high degree of maintenance to achieve best results; Heat and drought tolerance only reasonable, although grass will persist; Not greatly wear tolerant; Strong tendency to thatch.

COMMENTS Does not mix well with other species, but will become dominant; Commonly used where a high quality surface of fine grass is required; Requires a high level of maintenance to achieve best condition.

IDENTIFYING CHARACTERS Auricles – absent; Blades – tapering, flat, scabrous margins, ridged on upper surface, both surfaces glabrous; Collar – broad, continuous, glabrous; Ligule – membranous, up to 6 mm long, margins toothed; Sheath – round, hairless, split but overlapping margins, red-purplish tint; Vernation – rolled.

SOWING, GROWING AND MOWING

	New sowing rate		Oversowing rate		Stolonising rate		Mowing Heights	
	g/m²	lbs/1000 sq ft	g/m²	lbs/1000 sq ft	kg/100 m²	lbs/1000 sq ft	mm	inches
High	12	2.5	10	2.0	40	82	30	1.2
Low	3	.6	2	.4	20	41	2	.08

TOLERANCE

High temperature

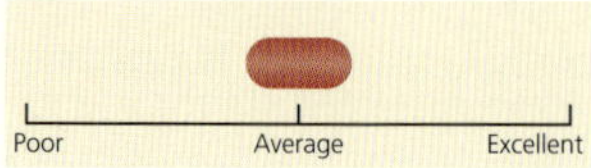

Drought

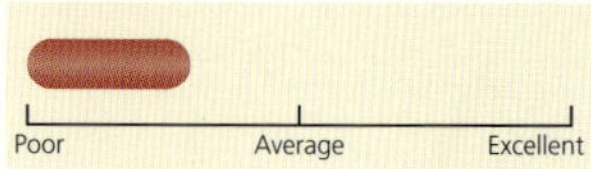

Close mowing

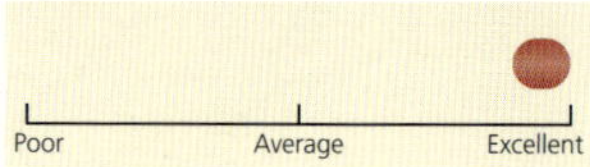

Low fertility

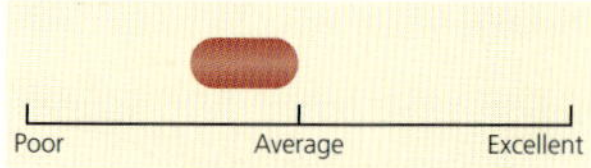

Frost

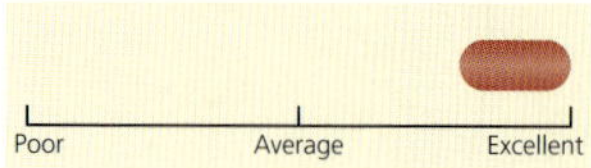

Shade

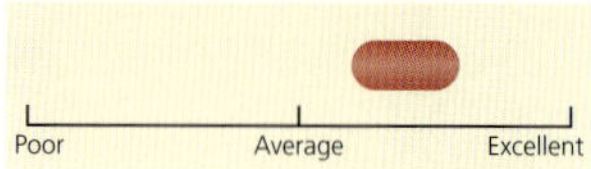

Wet soil

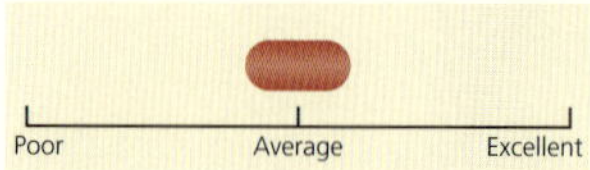

Wear

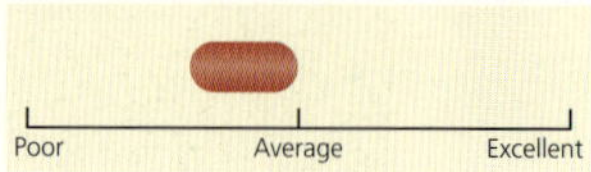

Salinity

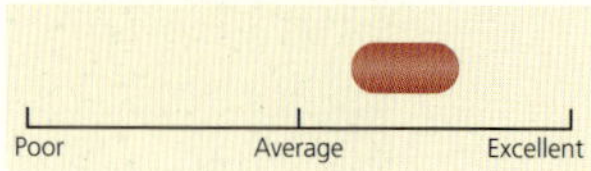

×9.6
×0.8

INCHES
CENTIMETRES

BOTANICAL NAME *Festuca arundinacea*

COMMON NAME Tall fescue

OTHER NAMES –

DESCRIPTION Texture variable – old cultivars have quite coarse leaves, while new cultivars are medium to fine; Colour usually a deep green, especially new cultivars; Density moderate to low; Deep and coarse root system; Bunch type grass, with some rhizomatous forms.

COMMON USES Winter sports fields, home lawns, polo fields, racecourses, lawn cemetries.

OTHER USES Golf tees, golf roughs.

POSITIVES Is quite heat and drought tolerant; Will withstand heavy wear once established; Retains colour and vigour under stress for longer than other cool season types; Will tolerate waterlogging for reasonable time; Some shade tolerance.

NEGATIVES If sown at low density or damaged, will form large clumps rather than uniform turf; Relatively slow to establish from seed.

COMMENTS Tall fescue has gained in usage quite considerably since 1990 as new cultivars have developed; Mixes well with Kentucky bluegrass, Perennial ryegrass and Bermuda grass in some circumstances; Needs to be sown at high rates.

IDENTIFYING CHARACTERS Auricle – claw-like; Blades – tapering, flat, wide, scabrous margins, glabrous surfaces, ridged upper surface; Collar – broad and divided, sometimes minute hairs; Ligule – membranous, truncated, entire, up to 2 mm long; Sheath – round, hairless, split with overlapping margins; Vernation – rolled.

SOWING, GROWING AND MOWING

	New sowing rate		Oversowing rate		Stolonising rate		Mowing Heights	
	g/m²	lbs/1000 sq ft	g/m²	lbs/1000 sq ft	kg/100 m²	lbs/1000 sq ft	mm	inches
High	50	10.2	40	8.2	0	0	60	2.4
Low	35	7.2	25	5.1	0	0	25	1.0

TOLERANCE

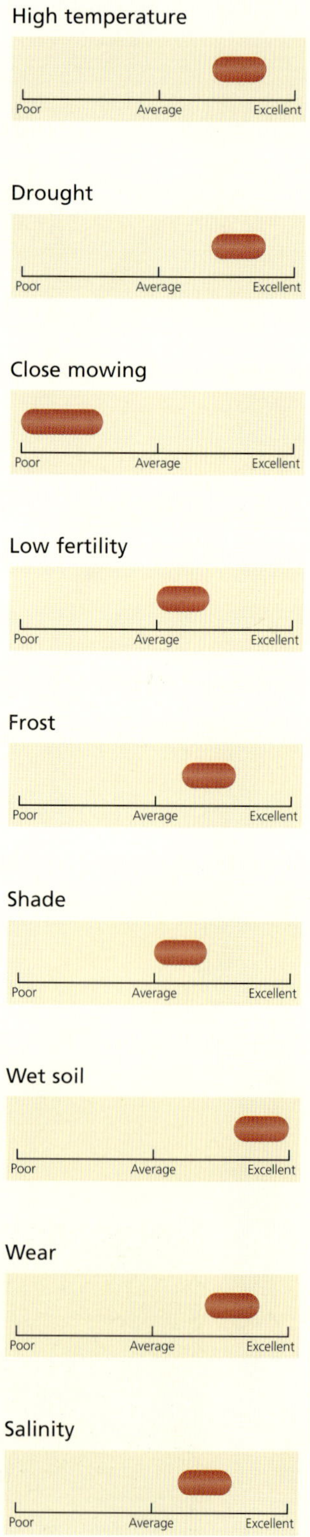

×4.0
×0.7

INCHES
CENTIMETRES

BOTANICAL NAME *Festuca longifolia*

COMMON NAME Hard fescue

OTHER NAMES –

DESCRIPTION Very fine textured grass that forms a full turf cover; Leaves are fine and bristle-like; Leaf colour is grey-green to green.

COMMON USES Golf surrounds, golf roughs, lawns, shaded lawns.

OTHER USES Low use areas that require grass cover with minimal maintenance.

POSITIVES This grass can be a low maintenance grass and still produce adequate turf quality; It can also produce high quality turf if maintained to a high degree; Reasonable shade tolerance; Can tolerate low fertility conditions.

NEGATIVES Slow establishment rate from seed; Can become thatchy if fertilised and not maintained regularly.

COMMENTS Widely used for shaded lawns, mixes well with other fine-leaf fescues for golf surrounds; Is highly persistent once established.

IDENTIFYING CHARACTERS: Auricles – absent; Blades – bristle-like, v-shaped, smooth to rough on upper surface, glabrous margin, grey to blue-green colour; Collar – divided, glabrous, narrow; Ligule – membranous, 0.1–0.3 mm long, short, not serrated or hairy; Sheath – very compressed, glabrous, red-brown at base; Vernation – folded.

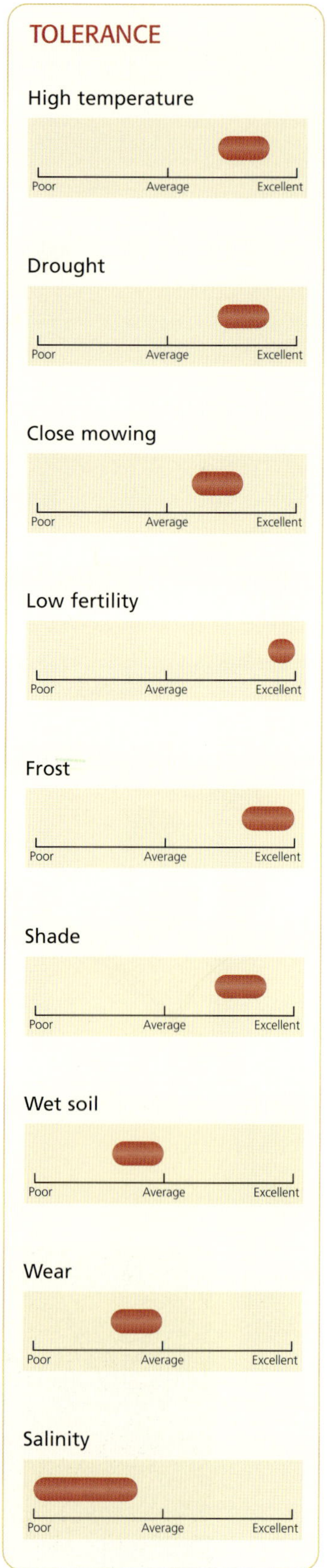

SOWING, GROWING AND MOWING

	New sowing rate		Oversowing rate		Stolonising rate		Mowing Heights	
	g/m^2	lbs/1000 sq ft	g/m^2	lbs/1000 sq ft	$kg/100\ m^2$	lbs/1000 sq ft	mm	inches
High	15	3.1	20	4.1	0	0	10	0.4
Low	5	1.0	10	2.0	0	0	20	0.8

×1.25

×30.0

BOTANICAL NAME *Festuca rubra* ssp. *commutata*

COMMON NAME Chewing's fescue

OTHER NAMES –

DESCRIPTION Very fine leaves forming a high density turf surface; Extensive tillering from bunch centre; Non-creeping habit distinguishes it from Red fescue types; Green to light green in colour; Can become bunchy if growing under poor conditions.

COMMON USES Golf tees and surrounds, fairway and greens oversowing, fine +/– shaded lawns.

OTHER USES Bowls/croquet, waste areas, parks and reserves.

POSITIVES Well adapted to shade; Tolerant of close mowing; Very high density gives very tight, uniform turf surfaces; Good drought tolerance.

NEGATIVES Not tolerant of heavy wear, particularly when wet; Will form excessive thatch in single grass stands unless frequently cultivated; Recuperation following damage is slow.

COMMENTS Has been used as a component of fine turf surfaces (e.g. golf greens); Commonly used as a component of seed mixtures to fill gaps; Extensive tillering and high density help to form excessive thatch.

IDENTIFYING CHARACTERS Auricles – absent; Blades – bristle like, v-shaped, tapering, glabrous margins; Collar – narrow, continuous, oblique; Ligule – membranous, small, up to 0.5 mm long, entire; Sheath – not compressed, minutely hairy, basal brown colour; Vernation – folded.

TOLERANCE

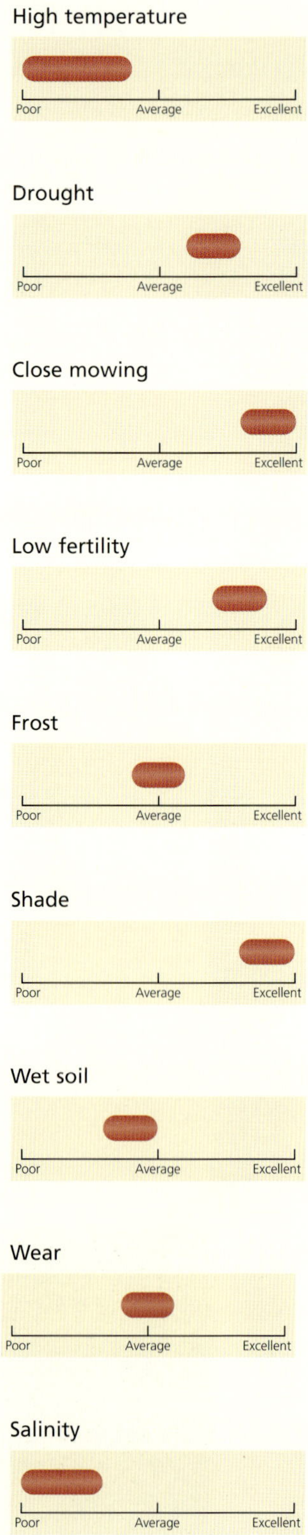

SOWING, GROWING AND MOWING

	New sowing rate		Oversowing rate		Stolonising rate		Mowing Heights	
	g/m²	lbs/1000 sq ft	g/m²	lbs/1000 sq ft	kg/100 m²	lbs/1000 sq ft	mm	inches
High	35	7.2	20	4.1	0	0	40	1.6
Low	25	5.1	15	3.1	0	0	2	.08

×1.4

×27

BOTANICAL NAME ***Festuca rubra* ssp. *rubra***

COMMON NAME Red fescue

OTHER NAMES Creeping red fescue

DESCRIPTION Very fine leaves and very high density; Colour from light green to green; Can be grown as an intensively managed fine turf; Creeping habit from rhizomes sets it apart from the Chewing's type fescues.

COMMON USES Golf tees and surrounds, fairway and greens oversowing, fine +/– shaded lawns.

OTHER USES Close mown sports, bowls/croquet, waste areas, parks and reserves

POSITIVES Can tolerate close mowing and will produce high density, high quality turf if well managed; Good shade tolerance; Creeping habit allows for some repair of divots; Blends well with Creeping bentgrass for fine turf appearance; Good drought tolerance.

NEGATIVES Tends to form excessive thatch; Wear tolerance only reasonable; Relatively slow to establish from seed.

COMMENTS Red fescue is used in mixtures to fill gaps which may occur in the turf cover; Good shade adaptation allows it to be used in lawns and other shaded areas; Can be used in fine turf because of its ability to tolerate close mowing.

IDENTIFYING CHARACTERS Auricles – absent; Blades – bristle-like, v-shaped, glabrous margins, hairy surface, deeply ridged on upper surface, smooth and slightly shiny on under surface; Collar – narrow, continuous, glabrous, pale green; Ligule – membranous, truncated, up to 0.5 mm long, entire; Sheath – not compressed, finely hairy, split part way only, reddish-brown at base; Vernation – folded.

SOWING, GROWING AND MOWING

	New sowing rate		Oversowing rate		Stolonising rate		Mowing Heights	
	g/m²	lbs/1000 sq ft	g/m²	lbs/1000 sq ft	kg/100 m²	lbs/1000 sq ft	mm	inches
High	35	7.2	20	4.1	0	0	50	2.0
Low	25	5.1	10	2.0	0	0	3	.1

TOLERANCE

High temperature

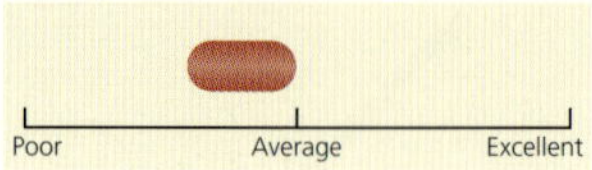

Drought

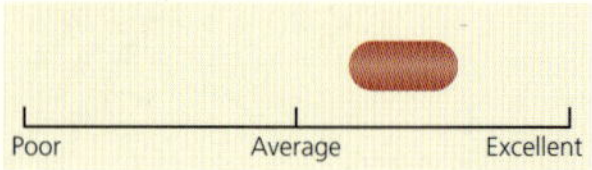

Close mowing

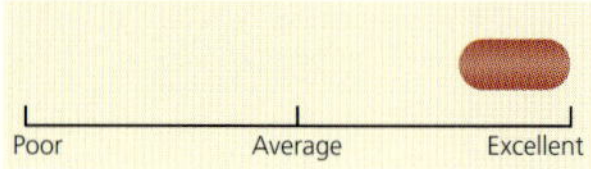

Low fertility

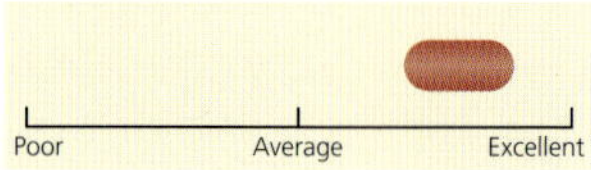

Frost

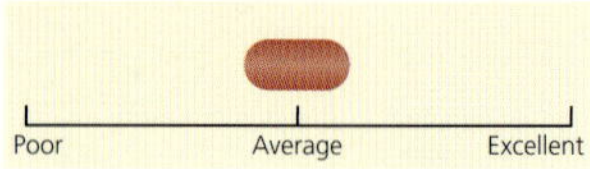

Shade

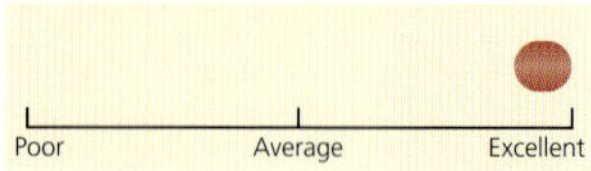

Wet soil

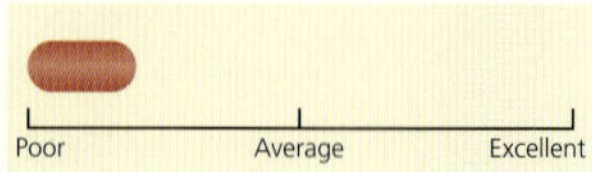

Wear

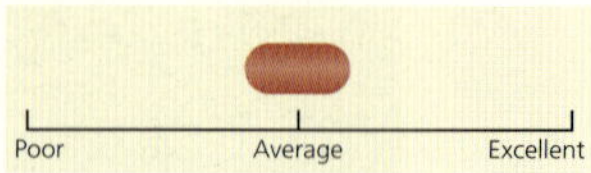

Salinity

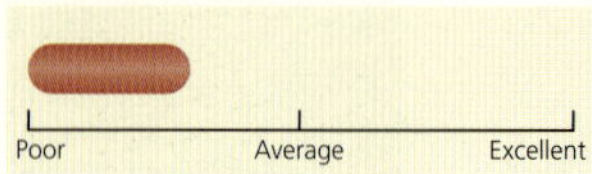

×20
×0.8

INCHES
CENTIMETRES

BOTANICAL NAME *Lolium perenne*

COMMON NAME Perennial ryegrass

OTHER NAMES Ray grass

DESCRIPTION Medium texture; Moderate shoot density; Leaves are glossy and normally emerald green in colour; A bunch type perennial grass with no lateral growth from stolons or rhizomes.

COMMON USES Golf tees, fairway oversowing, lawns, quality sport fields, winter sports fields, parks and reserves.

OTHER USES Wide usage where tolerance to wear and strong colour are required.

POSITIVES Very fast establishment from seed; Hard wearing when established; Will retain colour when frosted; Will tolerate wet soil conditions for some time; Can tolerate close mowing heights.

NEGATIVES Fast initial growth means that ryegrass can dominate seed mixtures; Not tolerant of salt, heat or drought; Requires regular nitrogen applications for best condition; Often can be difficult to mow; Susceptible to many grass diseases.

COMMENTS One of the most widely used turf grasses; Combines well with Kentucky bluegrass and Red fescue; Widely used for overseeding of Bermuda grass surfaces for winter vigour and colour; Excellent seedling vigour allows for sowing throughout the year.

IDENTIFYING CHARACTERS Auricles – variable, sometimes, long and clawlike, other times absent or rudimentary; Blades – sharply tapering to the point, keeled, prominently ridged on upper surface, smooth and glossy on lower surface, margins slightly scabrous; Collar – narrow, continuous, distinct, glabrous; Ligule – membranous, thin, 0.5– 2.0 mm long, toothed near the apex; Sheath – variably compressed, glabrous, closed or split margins, reddish at base; Vernation – folded.

SOWING, GROWING AND MOWING

	New sowing rate		Oversowing rate		Stolonising rate		Mowing Heights	
	g/m²	lbs/1000 sq ft	g/m²	lbs/1000 sq ft	kg/100 m²	lbs/1000 sq ft	mm	inches
High	40	8.2	30	6.1	0	0	50	2.0
Low	30	6.1	20	4.1	0	0	4	0.2

TOLERANCE

High temperature

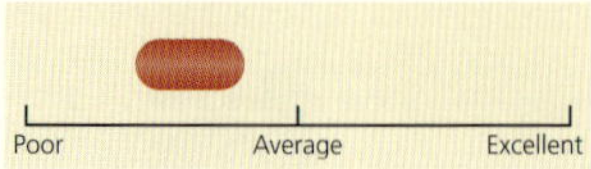

Drought

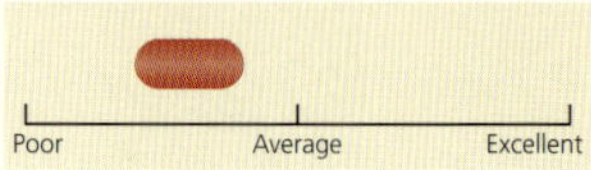

Close mowing

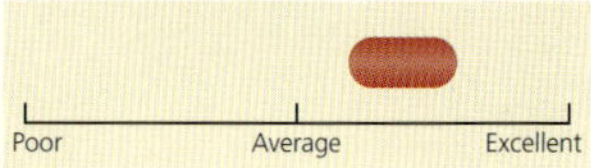

Low fertility

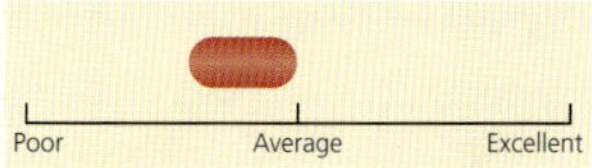

Frost

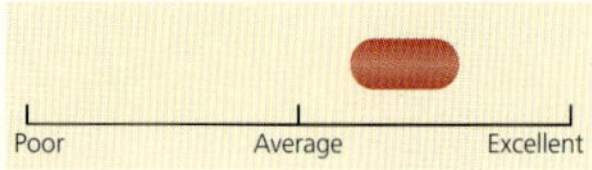

Shade

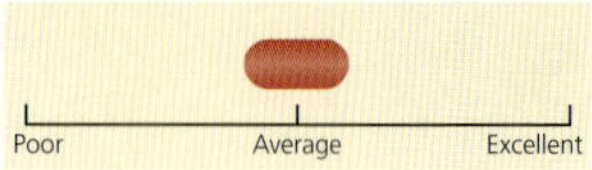

Wet soil

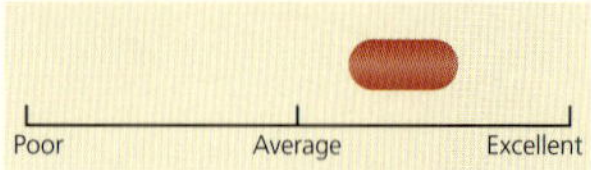

Wear

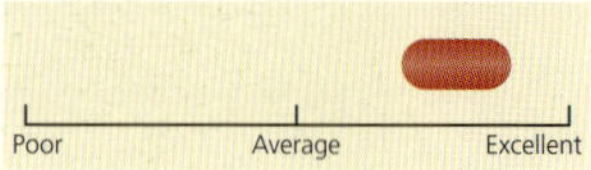

Salinity

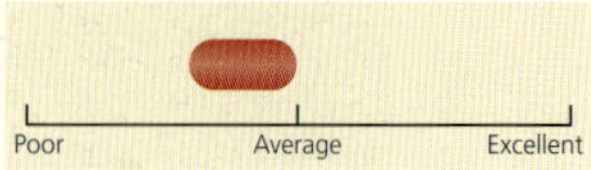

×0.8

×8.0

BOTANICAL NAME *Poa annua*

COMMON NAME Annual bluegrass

OTHER NAMES Winter grass, Poa, Annual poa

DESCRIPTION Small bunch type annual, with occasional stolons under close mowing; Leaves light green to yellowish green; Medium to fine texture depending on mowing height; Leaves not shiny on lower side.

COMMON USES Found in all temperate growing turf surfaces, commonly as a weed.

OTHER USES Not normally sown, but prevalent in many turfgrass surfaces.

POSITIVES Rapid germination from seed held in the topsoil and thatch layers; Can mix well with warm season grasses (e.g.Bermuda grass and Hybrid bermudagrass) to give good playing surface over the winter; Will quickly repair damaged areas; Requires moist conditions for growth.

NEGATIVES Annual bluegrass has a shallow root system and will not persist in dry periods; It is disease susceptible; Will not tolerate wear; Will form unsightly seed heads throughout much of the year; Strong thatching tendency; Profuse seeder.

COMMENTS Annual bluegrass is generally regarded as a weed of turf surfaces, but is included here since it forms such a high component of many turf surfaces; Seed is not widely available of any selected or improved types.

IDENTIFYING CHARACTERS Auricles – absent; Blades – flat or v-shaped, boat-shaped tip, often cross-wrinkled, margins glabrous, light green colour; Collar – glabrous, v-shaped broad continuous; Ligule – membranous, white, 1.0–2.0 mm long, acute, entire; Sheath – compressed, split part way only and overlapping, glabrous, pale green colour, slightly keeled; Vernation – folded.

SOWING, GROWING AND MOWING

	New sowing rate		Oversowing rate		Stolonising rate		Mowing Heights	
	g/m²	lbs/1000 sq ft	g/m²	lbs/1000 sq ft	kg/100 m²	lbs/1000 sq ft	mm	inches
High	40	8.2	0	0	0	0	30	1.2
Low	15	3.1	0	0	0	0	2	.08

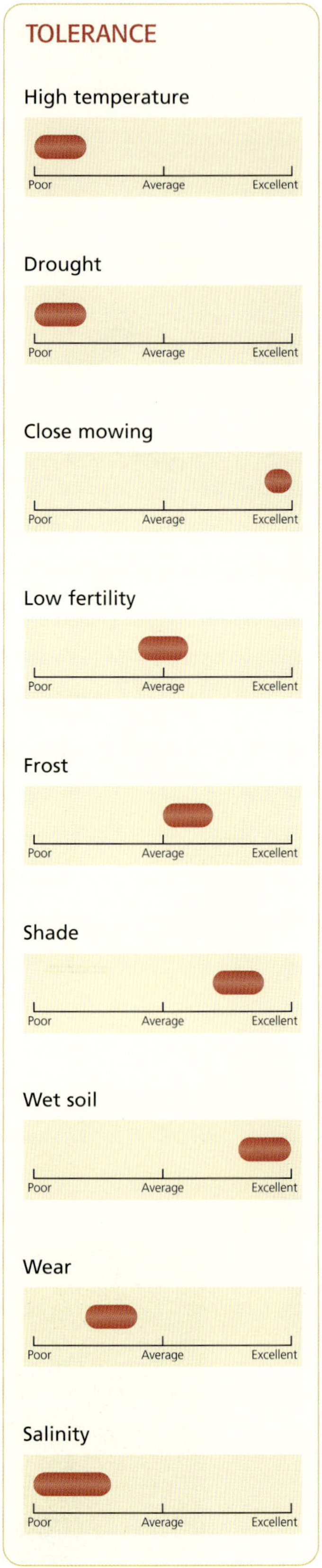

×1.0

×12.5

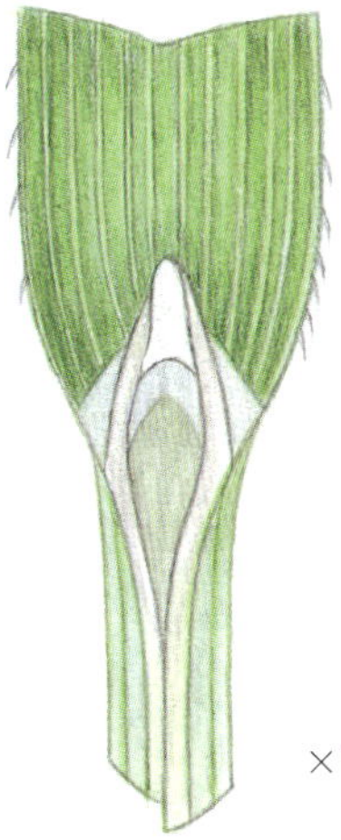

×15

BOTANICAL NAME *Poa pratensis*

COMMON NAME Kentucky bluegrass

OTHER NAMES Smooth stemmed meadow grass

DESCRIPTION Dark green colour; Medium texture; Moderate growth rate; Vigorous tillering; Shiny smooth leaves; Boat-shaped tip; Will form durable turf as develops numerous rhizomes; Hardwearing once established; Long lived grass in right location.

COMMON USES Golf fairways, tees and surrounds, lawns, quality sports fields, racecourses, parks and reserves, indoor sporting venues.

OTHER USES Forms a very durable and attractive turf in cool, temperate regions.

POSITIVES Is very persistent and durable once fully established; Colour and texture are attractive and soft; Can survive moderate drought stress; Will recuperate from damage by using its well-developed rhizome system.

NEGATIVES Slow to establish from seed; seedlings will not tolerate wear; Not known to tolerate low mowing, although some potential with new cultivars; Appears to need cool night temperatures for best growth.

COMMENTS Thatch can be a problem; Requires a high level of nitrogen; Frequently mixed with Perennial or Italian ryegrass and/or Tall fescue in instant turf blends; Good recuperative potential by means of rhizomes after damage.

IDENTIFYING CHARACTERS Auricles – absent; Blades – v-shaped, boat-shaped tip, sometimes shiny on under surface, margins scabrous and surfaces sometimes minutely pubescent; Collar – broad, divided, usually ciliate; Ligule – membranous, truncate, 0.2–0.6 mm long; Sheath – compressed, ciliate, split and sometimes overlapping, light green colour; Vernation – folded.

SOWING, GROWING AND MOWING

	New sowing rate		Oversowing rate		Stolonising rate		Mowing Heights	
	g/m²	lbs/1000 sq ft	g/m²	lbs/1000 sq ft	kg/100 m²	lbs/1000 sq ft	mm	inches
High	20	4.1	15	3.1	0	0	10	0.4
Low	10	2.1	8	1.6	0	0	20	0.8

TOLERANCE

High temperature

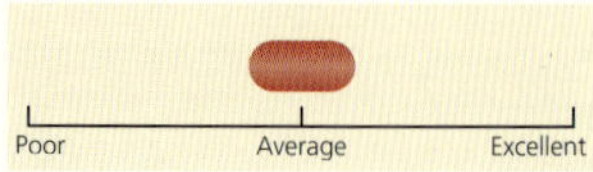

Drought

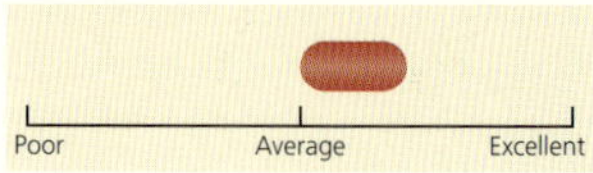

Close mowing

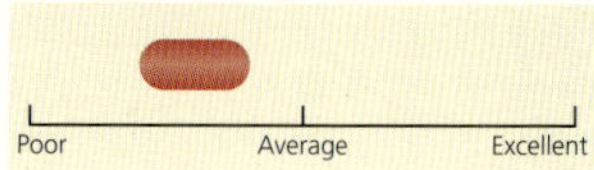

Low fertility

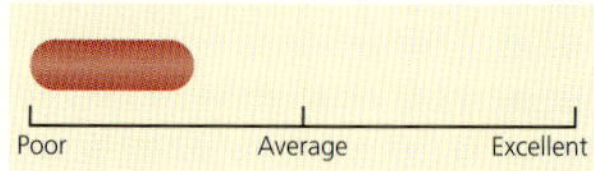

Frost

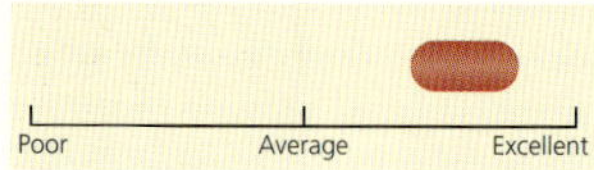

Shade

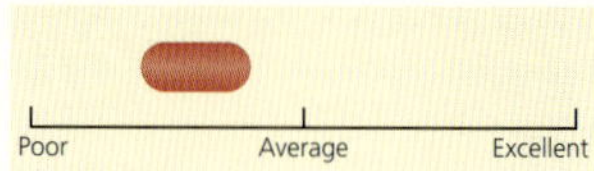

Wet soil

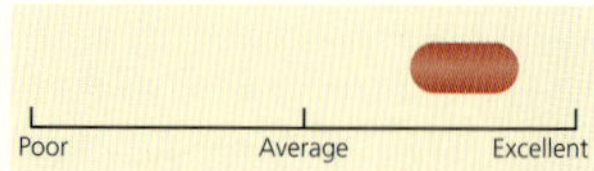

Wear

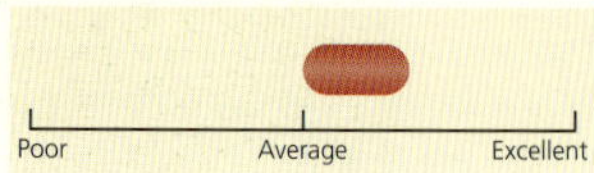

Salinity

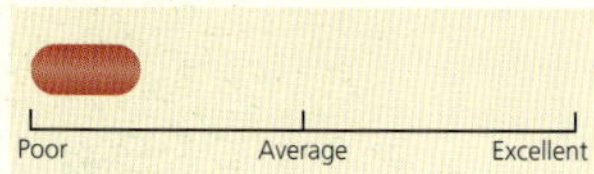

×12.5

×1.0

×12.5

BOTANICAL NAME *Poa trivialis*

COMMON NAME Rough stemmed bluegrass

OTHER NAMES Rough stalked meadow grass, Rough stemmed meadow grass

DESCRIPTION A fine textured grass appearing shiny on under surface of leaf; Yellowish-green leaves; Soft smooth leaf with a boat-shaped tip; Prostrate growth habit; Spreads by stolons.

COMMON USES Greens and fairway oversowing, lawns (particularly those in low light), golf aprons, lawn cemetries, parks and reserves.

OTHER USES Winter oversowing of warm-season turfs.

POSITIVES Prefers dry sterile soils (especially acidic soils) that cannot sustain Kentucky bluegrass; Seedling vigour is quite good compared with other Bluegrasses; Shade tolerance very good; Does not persist when used for oversowing into Bermuda and Hybrid bermudagrass areas.

NEGATIVES Shallow root system makes it susceptable to heat and drought stress; Some disease problems; May not tolerate some herbicides; Relatively low wear tolerance; Requires high fertiliser rates to develop a quality turf.

COMMENTS Used widely for overseeding of Bermuda grass fairways and greens for winter colour and early die-out in spring; Otherwise is commonly used in shaded lawns where cool and moist conditions prevail.

IDENTIFYING CHARACTERS Auricles – absent; Blades – tapering, flat, boat-shaped tip, slightly scabrous on upper surface, glossy and keeled on lower surface, margins scabrous; Collar – broad, glabrous, divided; Ligule – membranous, 2–3 mm long, acute, entire; Sheath – compressed, split part way only, generally scabrous, green or purple tinted; Vernation – folded.

SOWING, GROWING AND MOWING

	New sowing rate		Oversowing rate		Stolonising rate		Mowing Heights	
	g/m²	lbs/1000 sq ft	g/m²	lbs/1000 sq ft	kg/100 m²	lbs/1000 sq ft	mm	inches
High	20	4.1	15	3.1	0	0	30	1.2
Low	10	2.0	8	1.6	0	0	10	0.4

TOLERANCE

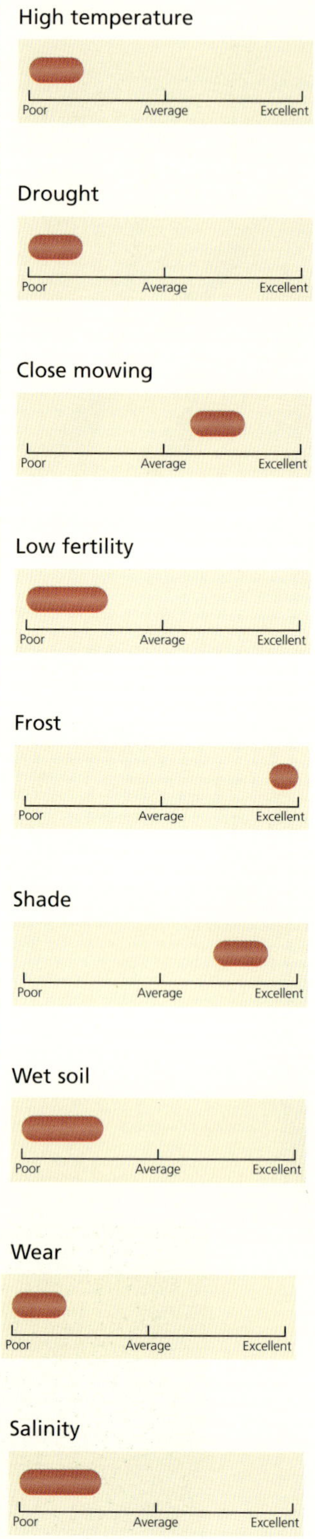

×15.0

×1.0

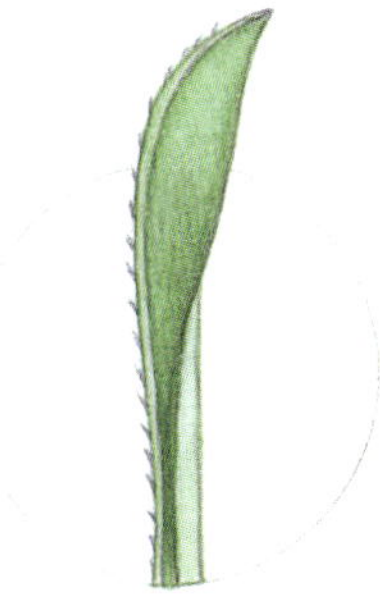

×12.5

BOTANICAL NAME *Axonopus affinis*

COMMON NAME Narrow-leaf carpetgrass

OTHER NAMES Common carpetgrass, Carpetgrass

DESCRIPTION Coarse textured leaf; Will form dense turf of moderate quality; Colour is green to light green in growing season becoming red when dormant under cool conditions; Prostrate growth habit with stoloniferous system of spreading.

COMMON USES Golf roughs, lawns, summer sports fields, sports fields.

OTHER USES Waste areas, parks and reserves.

POSITIVES Slow vertical growth rate means lower mowing requirement; Establishes quite well from seed; Will stabilise banks and batters quite quickly; Few thatch problems; Low fertility requirement; Makes best growth on sandy soils.

NEGATIVES Can be invasive with other grasses; Poor cold temperature tolerance, losing all colour and vigour; Does not tolerate close mowing; Poor wear tolerance; Lacks drought tolerance.

COMMENTS Ease of establishment from seed means that this grass is used extensively for revegetation of batters etc.; Generally not used where quality turf conditions are required; Can be propagated vegetatively; Low vertical growth rate.

IDENTIFYING CHARACTERS Auricles – absent; Blades – tapering, usually straight, smooth but sometimes hairs on upper surface, glabrous margins, leaf blade usually 5–10 mm wide; Collar – narrow, continuous, glabrous, not hairy at edges; Ligule – fringe of hairs; Sheath – very compressed, hairy, split with overlapping margins, red veins are obvious; Vernation – folded.

SOWING, GROWING AND MOWING

	New sowing rate		Oversowing rate		Stolonising rate		Mowing Heights	
	g/m²	lbs/1000 sq ft	g/m²	lbs/1000 sq ft	kg/100 m²	lbs/1000 sq ft	mm	inches
High	12	2.5	0	0	20	41	50	2.0
Low	8	1.6	0	0	5	10	20	0.8

TOLERANCE

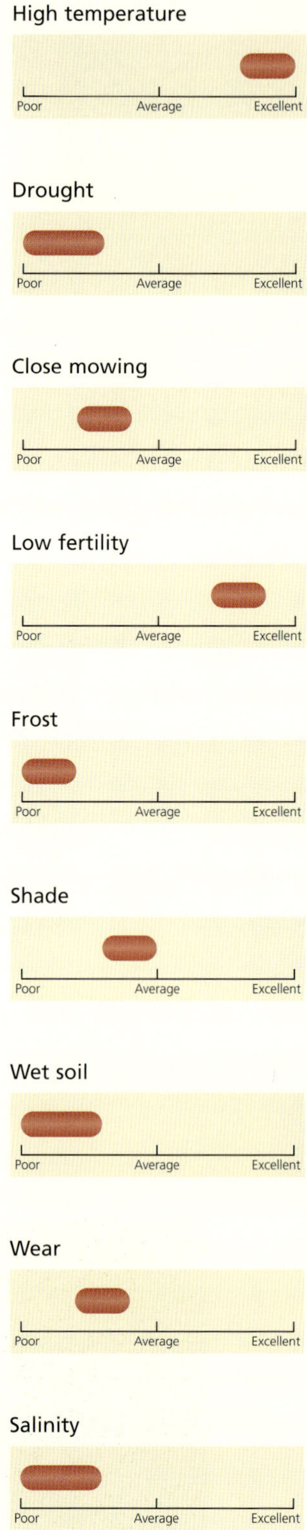

×0.8
×8

BOTANICAL NAME *Axonopus compressus*

COMMON NAME Broad-leaf carpet grass

OTHER NAMES Carpetgrass; Cow grass; Broad-leaf grass

DESCRIPTION Coarse textured grass that can form turf of low to moderate quality; Lustrous green colour and shiny leaves; Spreads by short rhizomes.

COMMON USES Golf roughs, passive recreation areas, racecourses, parks and reserves

OTHER USES Waste areas to stabilise soil.

POSITIVES Carpetgrass regenerates rapidly after damage; It has good colour; It is easy to establish from seed or from runners; Moderate shade tolerance and grows well under low fertility conditions.

NEGATIVES Coarse texture limits its usefulness to low quality areas or where texture is less important; Can sometimes be invasive of high quality turf when conditions are poor.

COMMENTS Very easy establishment from seed permits its use for revegetation of bare areas and for rapid repair of damaged turf; Generally not used where quality turf conditions are required; Can be propagated vegetatively if required.

IDENTIFYING CHARACTERS Auricles – absent; Blades – leaf blade wide (> 10 mm wide), often undulating, hairs on upper surface, glabrous margins; Collar – continuous, narrow, glabrous, not hairy at edges; Ligule – fringe of hairs; Sheath – very compressed, overlapping margins; Vernation – folded.

SOWING, GROWING AND MOWING

	New sowing rate		Oversowing rate		Stolonising rate		Mowing Heights	
	g/m²	lbs/1000 sq ft	g/m²	lbs/1000 sq ft	kg/100 m²	lbs/1000 sq ft	mm	inches
High	12	2.5	0	0	20	41	50	2.0
Low	8	1.6	0	0	5	10	20	0.8

TOLERANCE

High temperature

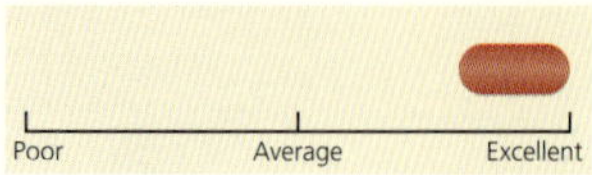

Drought

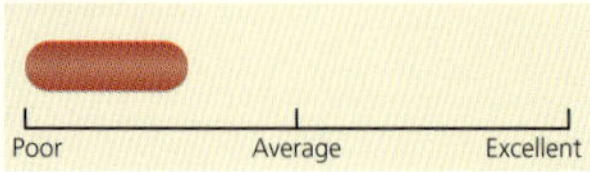

Close mowing

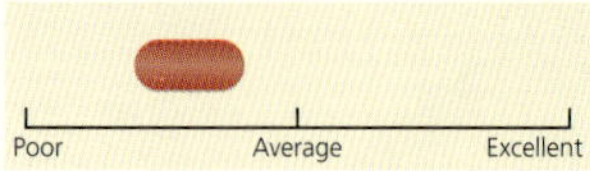

Low fertility

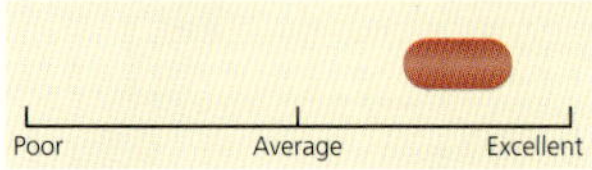

Frost

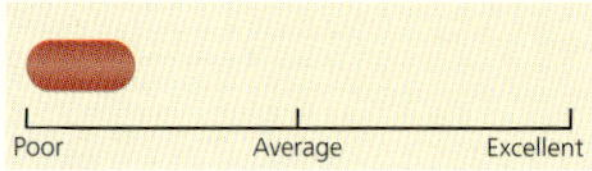

Shade

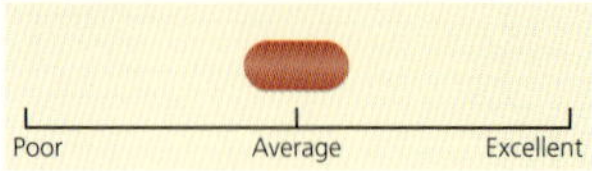

Wet soil

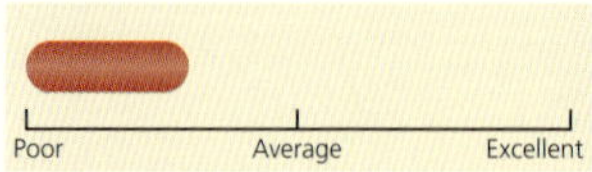

Wear

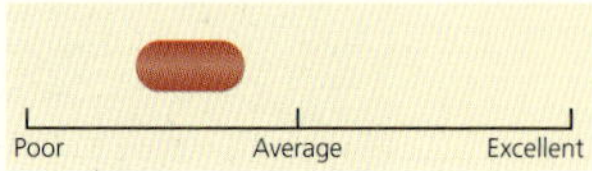

Salinity

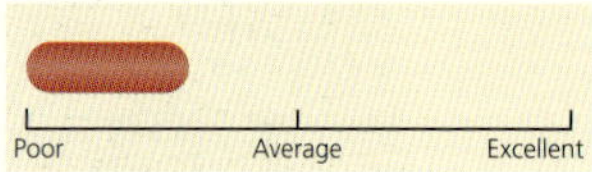

×4.0

×0.8

BOTANICAL NAME ***Buchloe dactyloides***

COMMON NAME Buffalo grass

OTHER NAMES US Buffalo grass

DESCRIPTION Fine textured grass of moderate density; Appearance often similar to drought stressed and uncut Bermuda and Hydrid bermudagrass; Prostrate growth habit from stolons; Colour grey-green to green; Will form a complete turf.

COMMON USES Golf roughs, lawns, low use sports fields, roadside areas, parks and reserves.

OTHER USES Waste areas, landfill areas.

POSITIVES Drought tolerance is very good; Low growth rate means less need for mowing; Will form a complete turf surface and bind soil together; Little thatching tendency; Low fertiliser requirement; Adapted to infrequent mowing and erosion control on heavy soils.

NEGATIVES Relatively slow to establish from seed; Poor wear tolerance; Seed heads form and can be unsightly.

COMMENTS Buffalo grass is often used where the capacity to mow an area is limited; It has a low growth rate and very good capacity to tolerate drought; Not normally mixed with other grass types; Produces both male and female plants.

IDENTIFYING CHARACTERS Auricles – absent; Blades – narrow leaf blades (< 5 mm), flat, sparsely ciliate, margins scabrous, grey-green colour; Collar – broad, continuous, glabrous; Ligule – fringe of hairs; Sheath – round, open, glabrous; Vernation – rolled.

SOWING, GROWING AND MOWING

	New sowing rate		Oversowing rate		Stolonising rate		Mowing Heights	
	g/m^2	lbs/1000 sq ft	g/m^2	lbs/1000 sq ft	kg/100 m^2	lbs/1000 sq ft	mm	inches
High	10	2.1	6	1.2	50	102	50	2.0
Low	5	1.0	3	0.6	20	41	15	0.6

TOLERANCE

High temperature

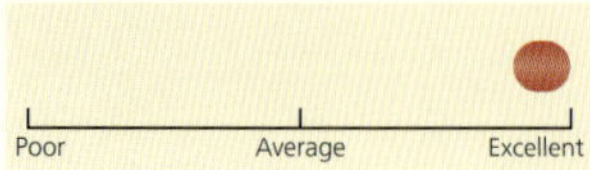

Drought

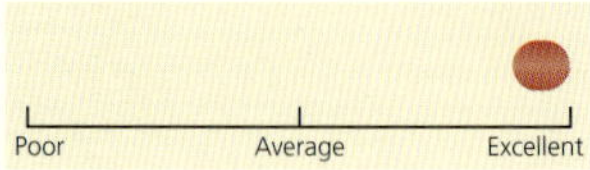

Close mowing

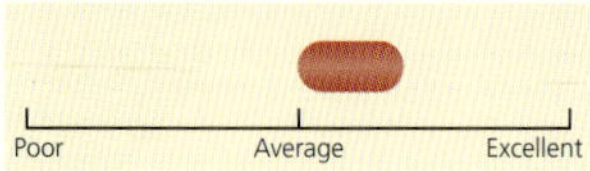

Low fertility

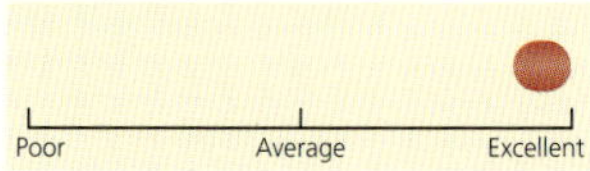

Frost

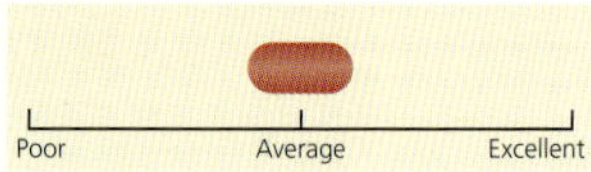

Shade

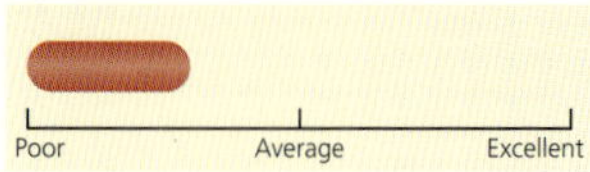

Wet soil

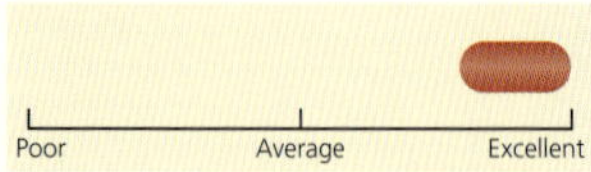

Wear

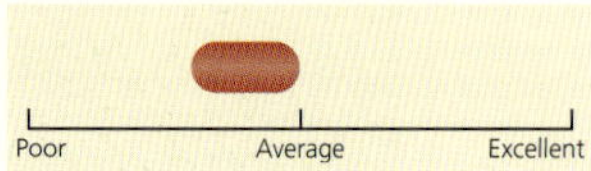

Salinity

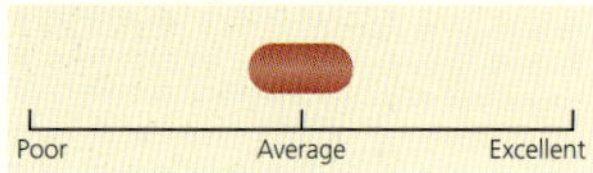

×16
×2.4
×1.2

INCHES
CENTIMETRES

BOTANICAL NAME *Cynodon dactylon*

COMMON NAME Bermuda grass

OTHER NAMES Couchgrass, Indian doub

DESCRIPTION Medium texture becoming finer under greater density; Prostrate growth habit by means of both rhizomes and stolons; Root system very fibrous and forms a tight sod.

COMMON USES Golf surrounds, fairways and tees, lawns, summer sports fields, bowls/croquet, parks and reserves, lawn cemeteries, racetracks

OTHER USES Wide adaptability because of fast growth and recovery.

POSITIVES Rapid growth rate and lateral spreading rates make this grass useful for areas which are frequently damaged or divoted; Drought and heat tolerant; Wear tolerant and somewhat salt tolerant; Recovery rate after wear very good.

NEGATIVES Poor cold weather tolerance, will lose colour and vigour when frosted; Poor growth in shade; Tendency to form deep thatch layer; Will invade nearby turf grass areas.

COMMENTS One of the most widely used turfgrasses with application in many warm and cool season environments; Can produce a fine turf when under a high level of maintenance; Capable of tolerating close mowing and hot, dry conditions; Not wear tolerant under cool conditions.

IDENTIFYING CHARACTERS Auricles – absent; Blades – mostly flat, stiff, sparsely pubescent on upper surface, usually glabrous on lower surface, margins scabrous, from 1.5 to 3 mm wide, hairy at base of leaf; Collar – continuous, medium width, glabrous, sparingly ciliate; Ligule – fringe of hairs; Sheath – compressed, sparsely pubescent, split with overlapping margins; Vernation – folded.

SOWING, GROWING AND MOWING

	New sowing rate		Oversowing rate		Stolonising rate		Mowing Heights	
	g/m²	lbs/1000 sq ft	g/m²	lbs/1000 sq ft	kg/100 m²	lbs/1000 sq ft	mm	inches
High	15	3.1	0	0	20	41	40	1.6
Low	10	2.1	0	0	5	10	4	0.2

TOLERANCE

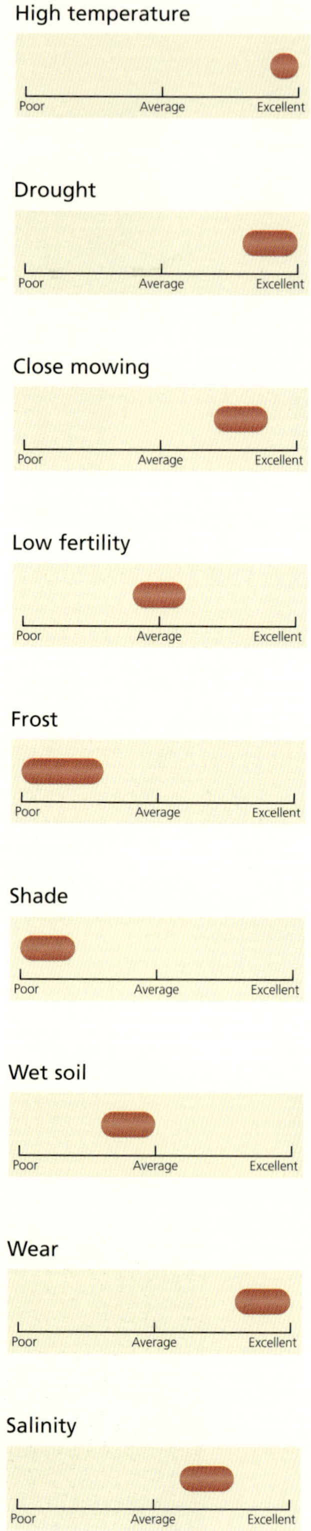

×0.64

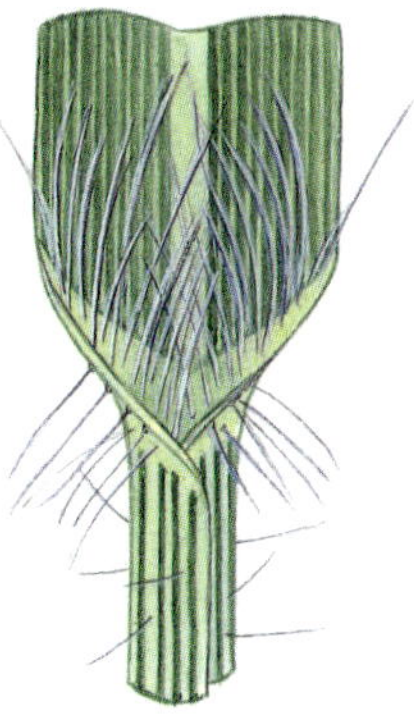

×8.0

BOTANICAL NAME *Cynodon dactylon* × *C. transvaalensis*

COMMON NAME Hybrid bermudagrass

CULTIVAR NAMES Hybrid bermudagrass cv, Tifgreen, Santa Ana, Tifdwarf, Tifeagle, Tifway, Champion, Floradwarf, Tifsport and others

DESCRIPTION Fine textured and high density turfgrasses; Prostrate growth from rhizomes and stolons; Colour generally dark green; Form seed heads but will not produce viable seed; Stolons root freely at the nodes; Deep root system.

COMMON USES Golf tees, fairways, surrounds and greens, cricket wickets.

OTHER USES Fine lawns, summer sports fields, quality sports fields.

POSITIVES Can tolerate close mowing and can form fine turf; Quite heat and drought tolerant; Very rapid growth and establishment rates from stolons; Will repair quickly from damage during growth period, good spring green up.

NEGATIVES Tends to form excessive thatch and requires regular de-thatching; Can only be propagated vegetatively; Susceptible to some Patch diseases; Winter vigour of some Bermudagrass hybrids is quite poor.

COMMENTS Widely used turfgrasses that originated from a cross between Common bermudagrass and South African Bermudagrass; All are propagated vegetatively; Does not mix well with other grasses; Has wide range of uses in warm climates.

IDENTIFYING CHARACTERS Auricles – absent; Blades – mostly flat, stiff, a few scattered hairs on upper surface, usually glabrous on lower surface, margins scabrous, from 1.0 to 2 mm wide; Collar – continuous, medium width, glabrous, sparingly ciliate; Ligule – fringe of hairs; Sheath – compressed, sparsely pubescent, split with overlapping margins; Vernation – folded.

SOWING, GROWING AND MOWING

	New sowing rate		Oversowing rate		Stolonising rate		Mowing Heights	
	g/m²	lbs/1000 sq ft	g/m²	lbs/1000 sq ft	kg/100 m²	lbs/1000 sq ft	mm	inches
High	0	0	0	0	20	41	30	1.2
Low	0	0	0	0	5	10	2	0.8

TOLERANCE

High temperature

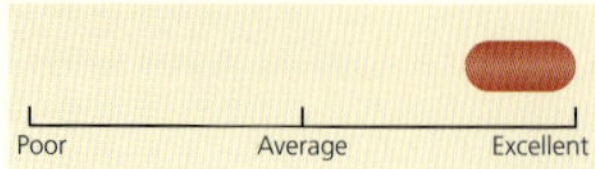

Drought

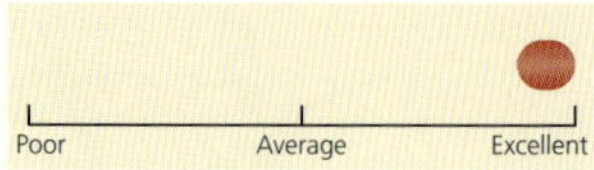

Close mowing

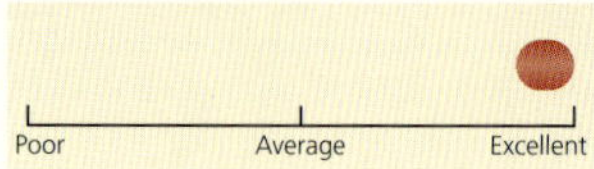

Low fertility

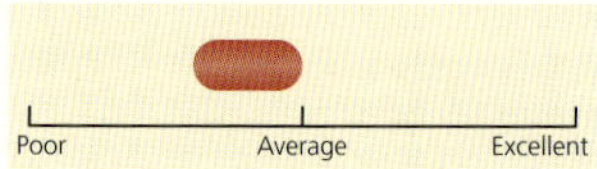

Frost

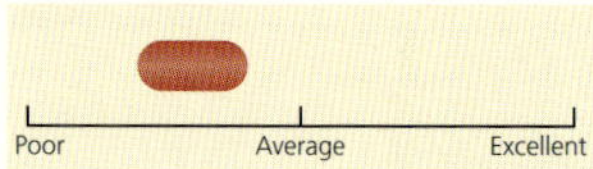

Shade

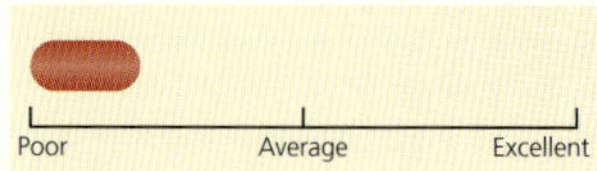

Wet soil

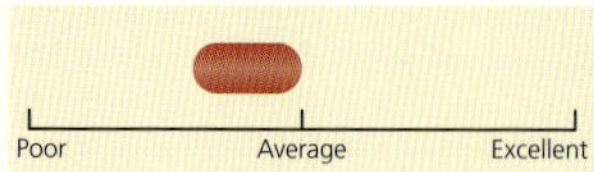

Wear

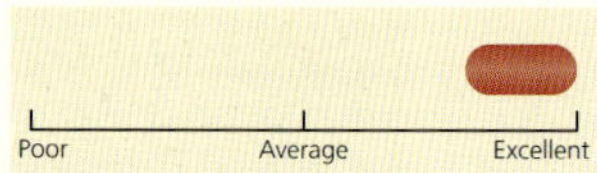

Salinity

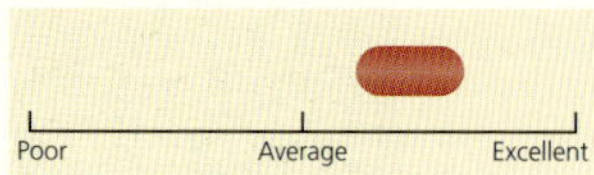

×1.5

×20

BOTANICAL NAME ***Dactyloctenium australe***

COMMON NAME Durban grass

OTHER NAMES Sweet smother grass

DESCRIPTION Short-statured perennial grass that spreads by stolons to form a full turf, lustrous green colour of leaves.

COMMON USES Golf roughs and coarse fairways, sports fields, lawns, shaded lawns.

OTHER USES Has been used for revegetation.

POSITIVES Relatively low maintenance requirements; Produces moderate quality turf; Can tolerate some frost, shade and drought conditions.

NEGATIVES Requires reasonably high fertility for best growth; Can produce seed heads over an extended period.

COMMENTS Of South African subtropical origin, used primarily for lawns in warm season and low-use areas in shaded environments.

IDENTIFYING CHARACTERS Auricles – absent; Blades – not hairy on upper surface, tapering, flat, hairy margins, moderately broad; Collar – narrow, continuous, glabrous, hairy at edges; Ligule – fringe of hairs; Sheath – very compressed, split and overlapping, glabrous; Vernation – folded.

SOWING, GROWING AND MOWING

	New sowing rate		Oversowing rate		Stolonising rate		Mowing Heights	
	g/m²	lbs/1000 sq ft	g/m²	lbs/1000 sq ft	kg/100 m²	lbs/1000 sq ft	mm	inches
High	0	0	0	0	0	0	12	0.5
Low	0	0	0	0	0	0	50	2.0

TOLERANCE

High temperature

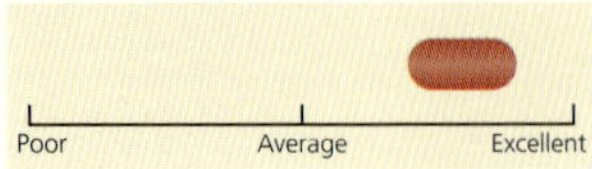

Drought

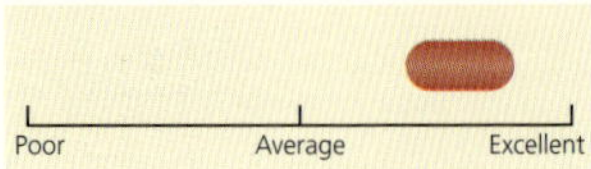

Close mowing

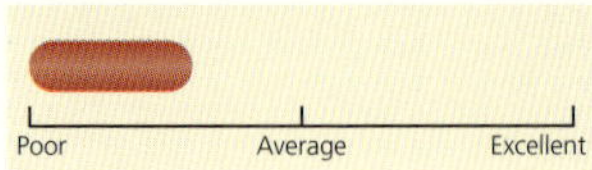

Low fertility

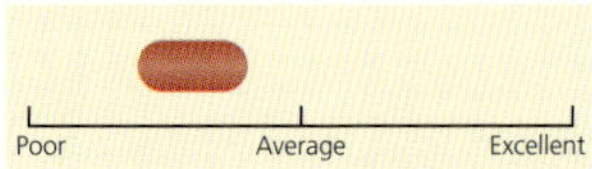

Frost

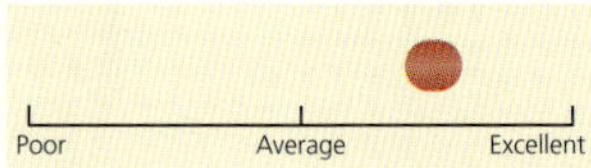

Shade

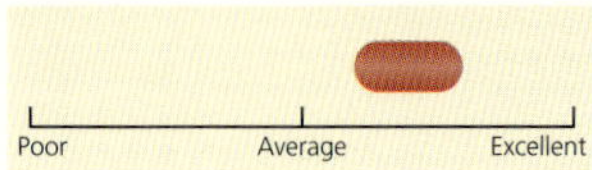

Wet soil

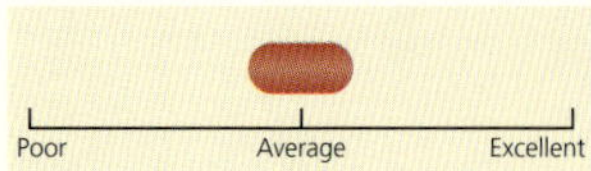

Wear

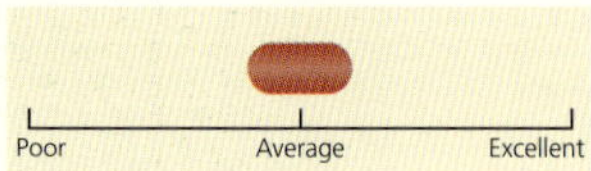

Salinity

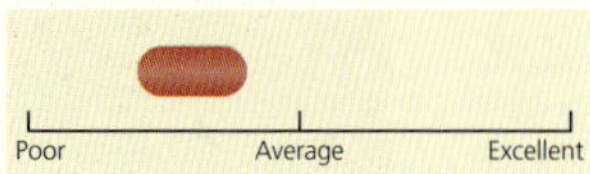

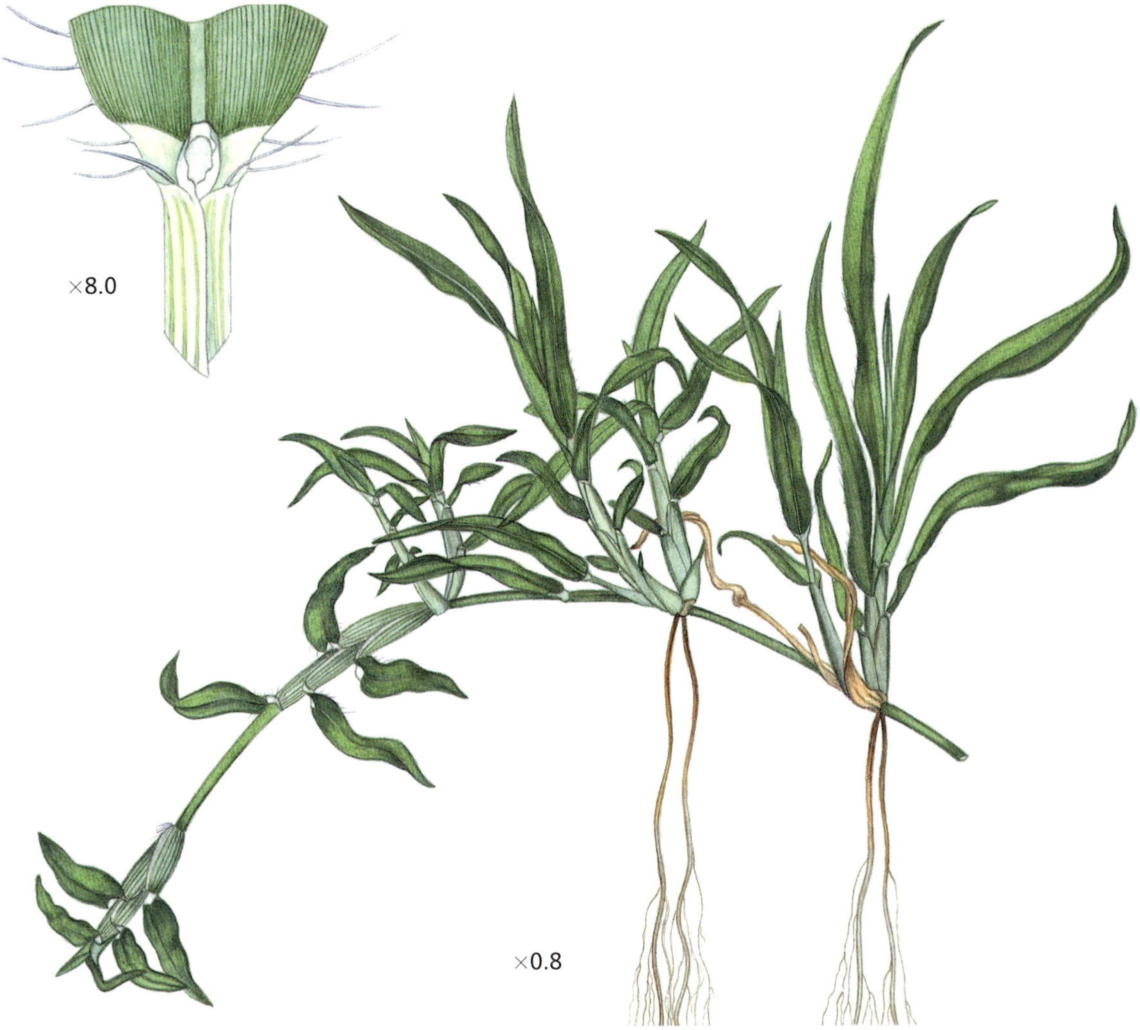

×8.0

×0.8

BOTANICAL NAME *Digitaria didactyla*

COMMON NAME Queensland blue couch

OTHER NAMES Serangoon grass, Blue couch

DESCRIPTION Medium to fine leaf texture; Density reasonable to slightly open; Prostrate growth by stolons; Colour blue-green to green; Will form complete turf surface; Loses all colour and vigour when frosted.

COMMON USES Golf tees, fairways, roughs and (sometimes) greens, lawns, sports fields, bowling greens.

OTHER USES Fine lawns, bowls/croquet, waste areas, parks and reserves.

POSITIVES Excellent colour and soft appearance make it suitable for lawns; Easy to grow from either seed or stolons.

NEGATIVES Poor cold weather tolerance; Only moderate drought tolerance; Somewhat poor tolerance of common broadleaf herbicides; Loses density when temperatures fall in winter.

COMMENTS Used intensively in south-eastern Queensland, central coast of Queensland and northern New South Wales, Australia and in South-East Asia; Mixes quite well with Tall fescue; Tolerates close mowing; Can be seeded or vegetatively propagated.

IDENTIFYING CHARACTERS Auricles – absent; Blades – tapering, glabrous margins, ciliate on both surfaces, bluish-green colour especially when stressed by heat or drought; Collar – broad, divided, few hairs at margins; Ligule – membranous, acute, entire, 2–3 mm long, often tinted at edges; Sheath – round, split with overlapping margins; Vernation – rolled.

SOWING, GROWING AND MOWING

	New sowing rate		Oversowing rate		Stolonising rate		Mowing Heights	
	g/m²	lbs/1000 sq ft	g/m²	lbs/1000 sq ft	kg/100 m²	lbs/1000 sq ft	mm	inches
High	12	2.5	10	2.0	30	61	25	1.0
Low	5	1.0	3	0.6	15	31	2	.08

TOLERANCE

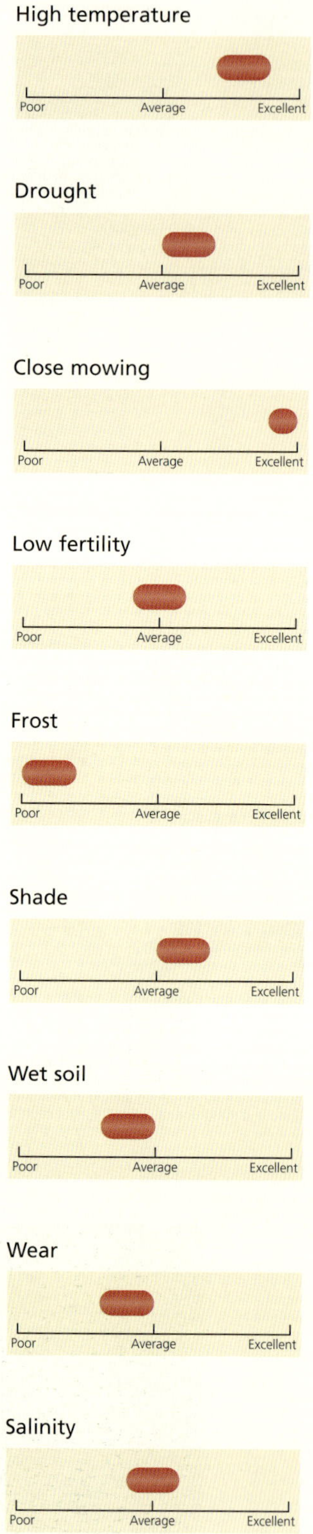

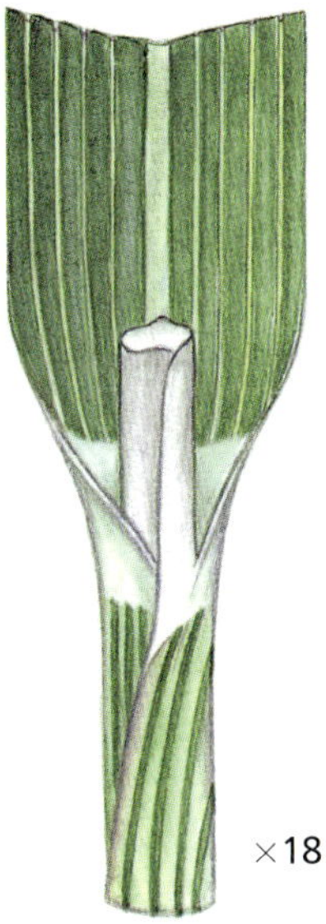

×18

×0.9

BOTANICAL NAME ***Eremochloa ophiuroides***

COMMON NAME Centipede grass

OTHER NAMES Chinese lawngrass

DESCRIPTION Medium textured grass with soft, almost succulent, leaves; Grows well under low traffic conditions; Shiny leaves and waxy feel; Colour light green to green.

COMMON USES Golf roughs, shaded lawns.

OTHER USES Relatively low maintenance areas.

POSITIVES Forms soft, luxurious lawn with high lustre; Relatively easy to establish from stolons; Some shade tolerance; Tolerant of acid soils.

NEGATIVES Can thatch badly if fertilised excessively; Poor wear tolerance; Not suited to cool, temperate growing conditions.

COMMENTS Not widely used because of relatively slow growth rate and slow repair after damage; Will form good turf under acidic soil conditions growing in association with low fertility.

IDENTIFYING CHARACTERS: Auricles – absent; Blades – flattened, 3–5 mm wide, blunt, keeled, glabrous under surface; Collar – continuous, broad, hairy; Ligule – short membrane with cilia; Sheath – very compressed, fused keel, glabrous but with greyish hairs at junction of leaf blade, margins overlapping, pale green; Vernation – folded.

TOLERANCE

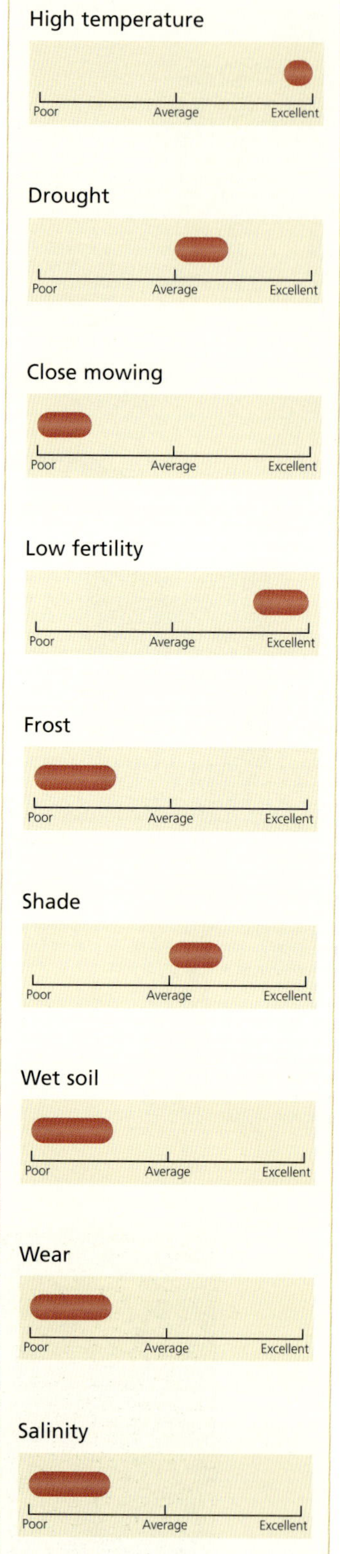

SOWING, GROWING AND MOWING

	New sowing rate		Oversowing rate		Stolonising rate		Mowing Heights	
	g/m²	lbs/1000 sq ft	g/m²	lbs/1000 sq ft	kg/100 m²	lbs/1000 sq ft	mm	inches
High	4	1	2	0.5	0	0	80	3.1
Low	2	0.5	1	0.25	0	0	15	0.6

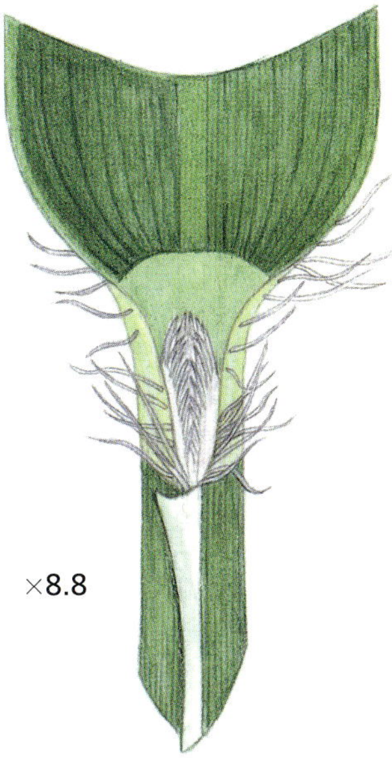

×8.8

×1.0

BOTANICAL NAME ***Paspalum dilatatum***

COMMON NAME Dallis grass

OTHER NAMES Paspalum

DESCRIPTION Long-lived perennial grass that spreads by short, fleshy rhizomes; Tends to colonise areas by slow spreading nature; Often has reddish tinge to leaf edges; Coarse to medium texture.

COMMON USES Summer sports fields, lawns, passive recreation areas.

OTHER USES Golf roughs, waste areas, parks and reserves

POSITIVES Very persistent once established; Forms complete surface if well maintained; Very tolerant of high temperatures; Wears well.

NEGATIVES Seedheads are unsightly and produce sticky resin; Plants are generally of low density; Grass is often considered a weed and is selectively removed.

COMMENTS Often considered a weed, this grass is useful in low-use surfaces that are required to produce a moderate quality of turf; Highly persistent under most turf conditions but can become dominant at certain times of year; Seedheads undesirable.

IDENTIFYING CHARACTERS Auricles – absent; Blades – crimped, reddish margins, glabrous on both surfaces; Collar – broad, continuous, hairy; Ligule – membranous, entire, acute; Sheath – slightly compressed or rounded, usually hairless, reddish-brown colour at base; Vernation – rolled.

SOWING, GROWING AND MOWING

	New sowing rate		Oversowing rate		Stolonising rate		Mowing Heights	
	g/m²	lbs/1000 sq ft	g/m²	lbs/1000 sq ft	kg/100 m²	lbs/1000 sq ft	mm	inches
High	5	1.0	0	0	0	0	80	3.1
Low	1	0.2	0	0	0	0	12	0.5

TOLERANCE

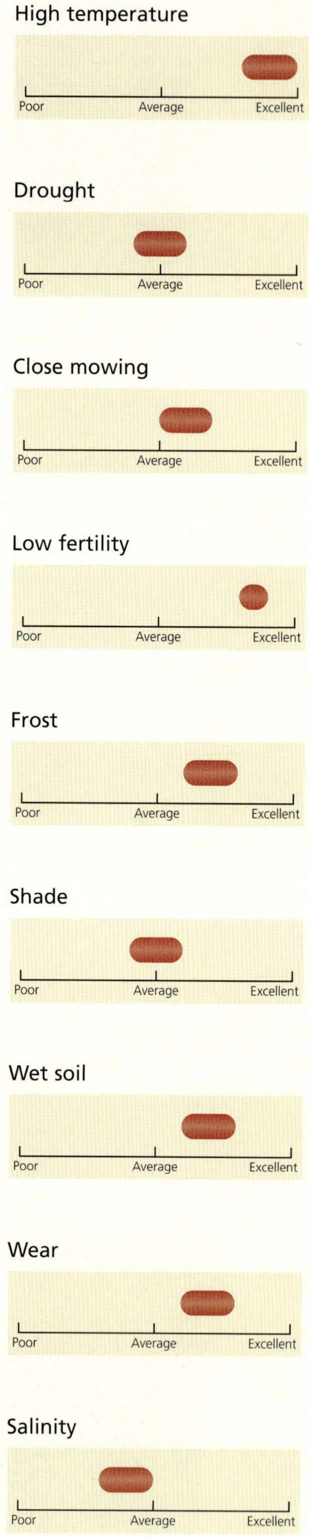

×0.8

×4.0

BOTANICAL NAME *Paspalum notatum*

COMMON NAME Bahia grass

OTHER NAMES –

DESCRIPTION A coarse, spreading grass that forms a full cover from an extensive, but shallow, rhizomateous system; Texture is coarse but leaves are quite soft; Several cultivars have been developed specifically for turf and amenity purposes.

COMMON USES Golf roughs, racecourses, low quality sports fields.

OTHER USES Waste areas, low use areas.

POSITIVES Will repair itself after damage from its shallow rhizome system; Forms turf of moderate density; Good drought tolerance; Tolerates regular mowing and occasional submersion.

NEGATIVES Not easy to establish from seed nor easy to mow; poor shade tolerance.

COMMENTS The creeping habit of this grass, through shallow rhizomes and/or stolons, and its strong root system mean that this grass can be a strong competitor once established; Slow initial growth and coarse texture limit its application.

IDENTIFYING CHARACTERS Auricles – absent; Blades – flat, margins ciliate towards base, broad to 15 mm wide, green, tapering, glabrous; Collar – glabrous, broad, continuous; Ligule – membranous, very blunt, entire; Sheath – slightly compressed, sometimes hairy, keeled, overlapping; Vernation – rolled.

SOWING, GROWING AND MOWING

	New sowing rate		Oversowing rate		Stolonising rate		Mowing Heights	
	g/m²	lbs/1000 sq ft	g/m²	lbs/1000 sq ft	kg/100 m²	lbs/1000 sq ft	mm	inches
High	5	1.0	0	0	0	0	15	0.6
Low	2	0.4	0	0	0	0	50	2.0

TOLERANCE

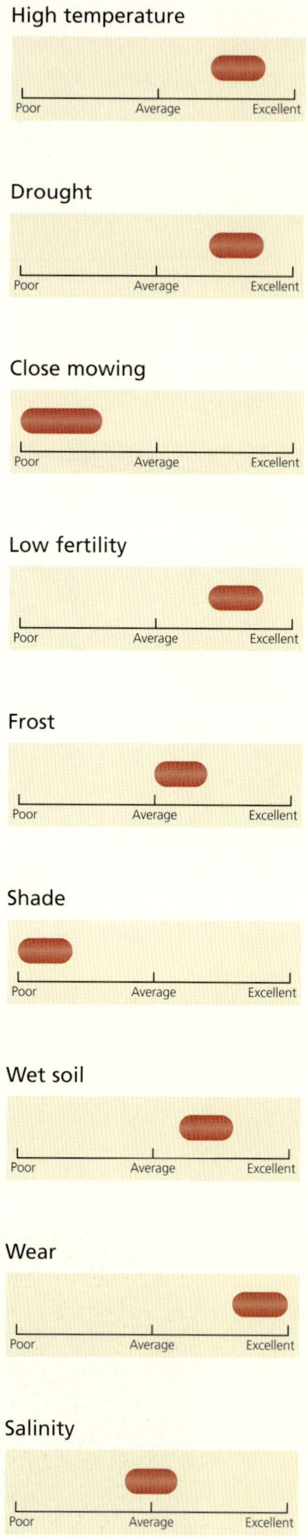

×8.0
×0.8

BOTANICAL NAME ***Paspalum vaginatum***

COMMON NAME Seashore paspalum

OTHER NAMES Saltwater couch, Siltgrass, Sandknotgrass

DESCRIPTION Fine to medium leaf texture that forms a full turf surface by way of stolons and rhizomes; Colour is green to dark green; Texture is soft and lush.

COMMON USES Golf tees, surrounds and fairways, summer sports fields, lawns.

OTHER USES Some new types may suit golf greens, saline prone areas.

POSITIVES Forms high quality turf under saline or non-saline conditions; Vigorous growth rate provides rapid coverage of new turf and shows good repair from wear; Good heat and drought tolerance.

NEGATIVES Relatively susceptible to disease and insect attack if badly managed; Selective weed control using chemicals is somewhat difficult; Can form excessive thatch if not maintained appropriately.

COMMENTS Aggressive lateral growth under both saline and non-saline growing conditions; Newer varieties can tolerate sea water as an irrigation source.

IDENTIFYING CHARACTERS Auricles – absent; Blades – tapering, v-shaped to flat, scabrous margins, often twisted, scattered hairs on upper surface, few hairs on lower surface; Collar – narrow, continuous, small tufts of hair; Ligule – membranous, short, entire; Sheath – slightly compressed, glabrous, red-brown colour; Vernation – rolled.

SOWING, GROWING AND MOWING

	New sowing rate		Oversowing rate		Stolonising rate		Mowing Heights	
	g/m²	lbs/1000 sq ft	g/m²	lbs/1000 sq ft	kg/100 m²	lbs/1000 sq ft	mm	inches
High	0	0	0	0	20	41	40	1.6
Low	0	0	0	0	5	10	10	0.4

TOLERANCE

High temperature

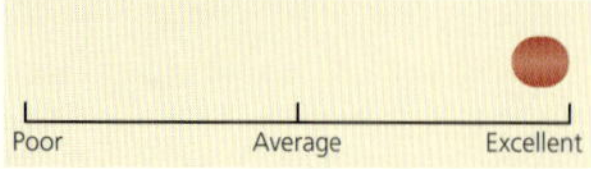

Drought

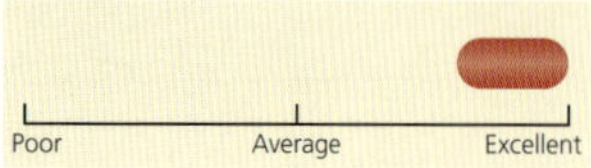

Close mowing

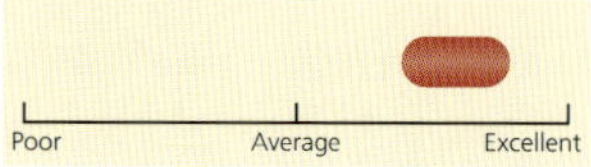

Low fertility

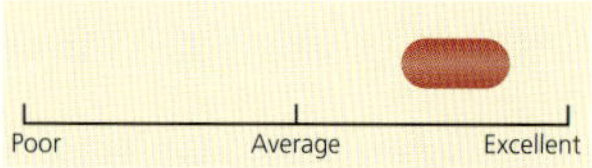

Frost

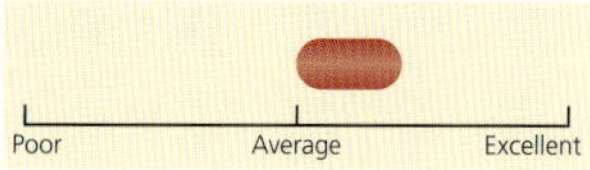

Shade

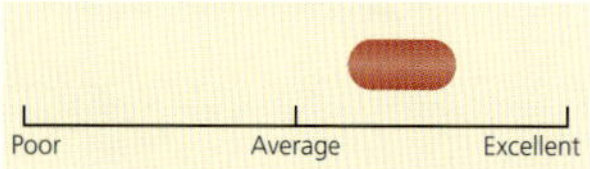

Wet soil

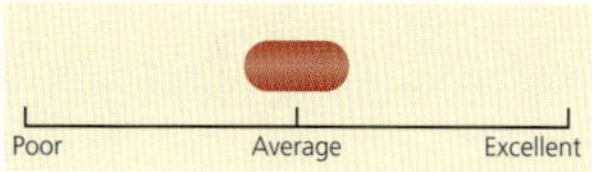

Wear

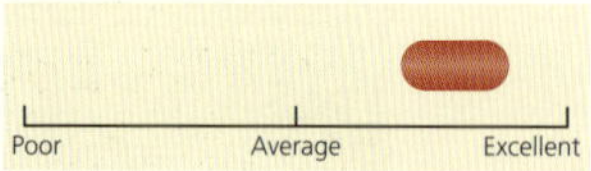

Salinity

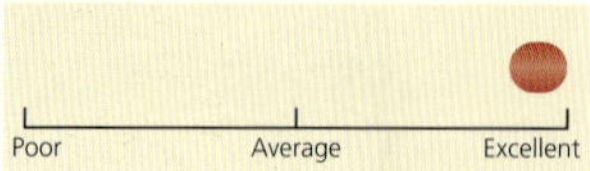

×10

×1.0

BOTANICAL NAME *Pennisetum clandestinum*

COMMON NAME Kikuyu grass

OTHER NAMES Kikuyo, Kikoejoe

DESCRIPTION Stoloniferous and rhizomatous perennial with very coarse textured leaves; Leaves yellow green; Forms thick dense turf which can become thatchy; Very aggressive nature with capacity to climb.

COMMON USES Golf tees and fairways, lawns, summer sports fields, racecourses, parks and reserves.

OTHER USES Waste and erosion prone areas.

POSITIVES Strong recuperative capacity after damage; Wear and drought tolerant; Easy to propagate from stolons; Useful in binding soil together after earthworks.

NEGATIVES Very invasive of other grasses, with rhizomes capable of surfacing metres away from the mother plant; Poor cold weather tolerance; Very strong thatching tendency; Not always easy to cut.

COMMENTS Considered a weed in many countries because of its invasive nature; Kikuyu grass can form an acceptable turf if mowed close and frequently dethatched; Difficult to control once established; Does not blend well with other grasses.

IDENTIFYING CHARACTERS Auricles – absent; Blades – hairy, flat, keeled, wide 4–5 mm, tapering point, ciliate margins, pale green colour; Collar – continuous, broad, hairy; Ligule – fringe of hairs; Sheath – very compressed, very hairy, split with overlapping margins; Vernation – folded.

SOWING, GROWING AND MOWING

	New sowing rate		Oversowing rate		Stolonising rate		Mowing Heights	
	g/m^2	lbs/1000 sq ft	g/m^2	lbs/1000 sq ft	kg/100 m^2	lbs/1000 sq ft	mm	inches
High	2	0.4	0	0	15	31	40	1.6
Low	1	0.2	0	0	5	10	8	0.3

TOLERANCE

High temperature

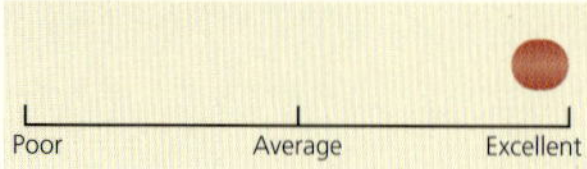

Drought

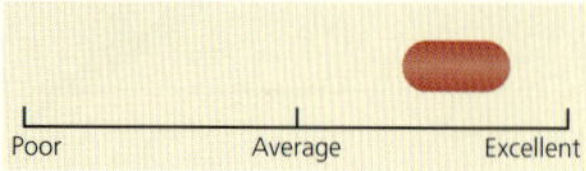

Close mowing

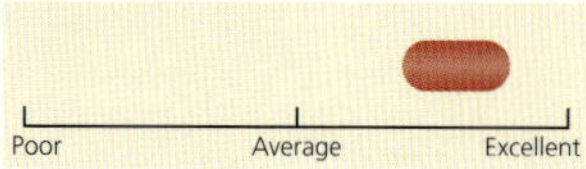

Low fertility

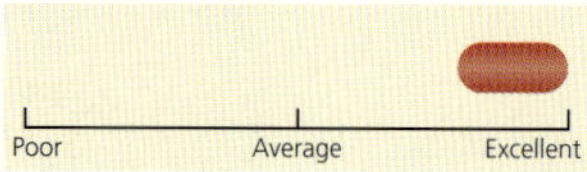

Frost

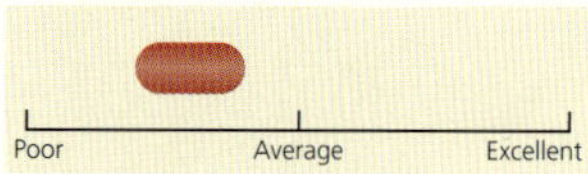

Shade

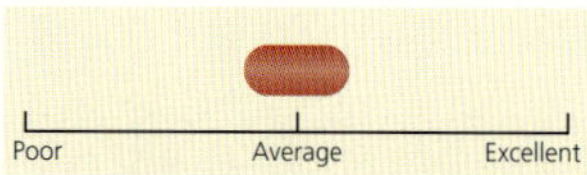

Wet soil

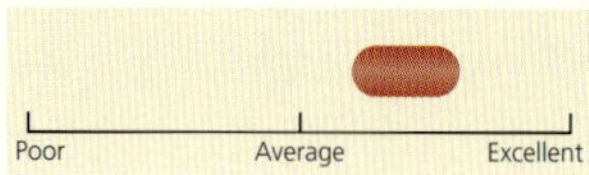

Wear

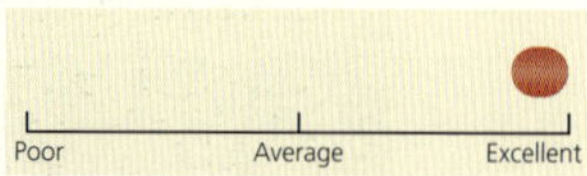

Salinity

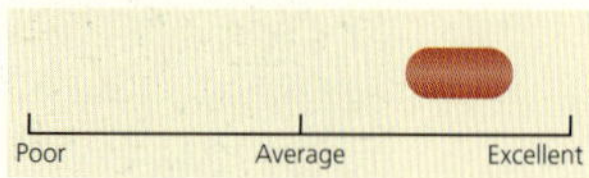

×4.8
×0.4

INCHES
CENTIMETRES

BOTANICAL NAME ***Stenotaphrum secundatum***

COMMON NAME St Augustine grass

OTHER NAMES Australian Buffalo grass

DESCRIPTION Very coarse texture; Blue-green colour; Prostrate growth from stolons; Fibrous root system; Forms a complete turf of medium density; Older varieties are not smooth, but slightly sharp-edged to walk upon or touch; Can form a thick thatchy turf which is hard to cut.

COMMON USES Lawns, shaded lawns, golf roughs.

OTHER USES Lightly shaded areas.

POSITIVES Good shade adaptation; Relatively low growth rate and nitrogen requirement; Reasonably drought hardy.

NEGATIVES Establishment rate from stolons is not fast except in very warm climates; Is susceptible to several diseases; Strong thatching tendency; Not tolerant to many common broadleaf herbicides; Can cause an allergic reaction in some people.

COMMENTS This grass is propagated by vegetative means since it produces very little viable seed; Very poor low temperature hardiness and changes colour in winter; It recuperates quite well after wear; Best grown on fertile, well-drained soils.

IDENTIFYING CHARACTERS Auricles – absent; Blades – usually flat, petioled, smooth, glabrous surface, blunt tip; Collar – continuous, broad, smooth; Ligule – inconspicuous fringe of hairs; Sheath – very compressed, keeled, slightly ciliate towards tip and along the margins; Vernation – folded.

SOWING, GROWING AND MOWING

	New sowing rate		Oversowing rate		Stolonising rate		Mowing Heights	
	g/m²	lbs/1000 sq ft	g/m²	lbs/1000 sq ft	kg/100 m²	lbs/1000 sq ft	mm	inches
High	0	0	0	0	40	82	40	1.6
Low	0	0	0	0	10	21	20	0.8

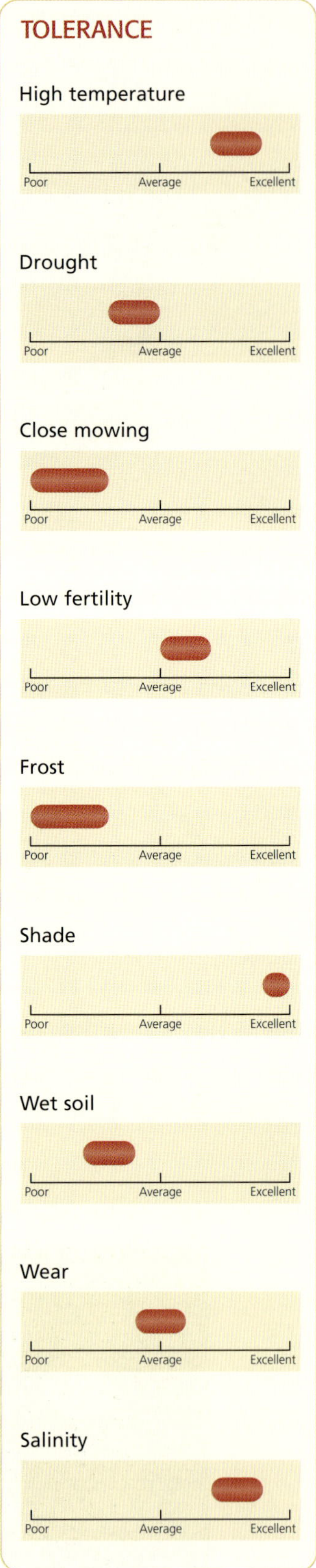

×2.4

×0.8

BOTANICAL NAME *Zoysia japonica*

COMMON NAME Japanese lawngrass

OTHER NAMES Zoysia

DESCRIPTION Medium to fine texture; Dense sward of prostrate growing plants with growth coming from rhizomes and stolons; Leaves are tough and stiff forming a very tight, durable surface; Colour light green to green; Produces seed, but generally planted vegetatively.

COMMON USES Golf fairways, tees and roughs, lawns, shaded lawns, summer sports fields, indoor grass sporting venues.

OTHER USES Waste areas, general-purpose turf and erosion prone areas.

POSITIVES Forms a tight turf of high quality when grown under correct conditions; Growth rate is quite low, requiring limited mowing; Good heat and drought tolerance; Very wear tolerant; Good salt tolerance; Grows best in heavy soils.

NEGATIVES Slow lateral growth means Japanese lawngrass is slow to establish and repair after damage; Winter colour and vigour are limited; Can form excessive thatch; In practice, establishment is likely to be only from turf (complete sods).

COMMENTS The use of Japanese lawngrass has been limited until recently; Poor cool season growth and recuperation also limit usage in transition areas; Does not mix well with other grasses; May require winter oversowing to sustain a surface; Hybrids are becoming more available.

IDENTIFYING CHARACTERS Auricles – absent; Blades – flat, stiff, sparsely hairy above and near base, smooth undersurface, glabrous margins, width usually 2–5 mm; Collar – broad, continuous, hairy on margins; Ligule – fringe of hairs; Sheath – slightly compressed, tufted hairs at junction with blade, split with overlapping margins; Vernation – rolled.

SOWING, GROWING AND MOWING

	New sowing rate		Oversowing rate		Stolonising rate		Mowing Heights	
	g/m²	lbs/1000 sq ft	g/m²	lbs/1000 sq ft	kg/100 m²	lbs/1000 sq ft	mm	inches
High	15	3.0	0	0	70	143	25	1.0
Low	10	2.0	0	0	30	61	10	0.4

TOLERANCE

High temperature

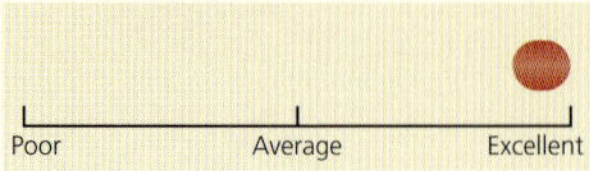

Drought

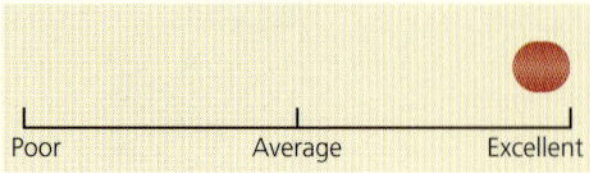

Close mowing

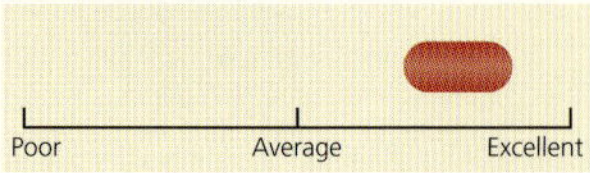

Low fertility

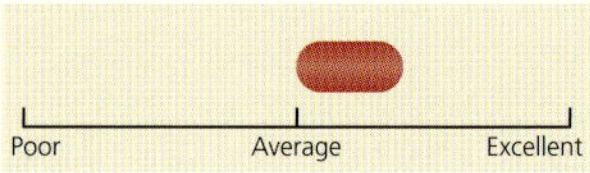

Frost

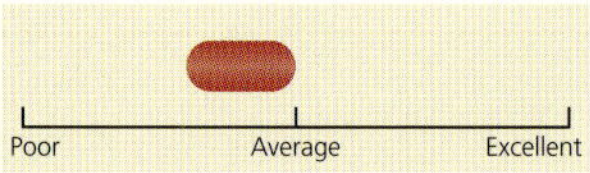

Shade

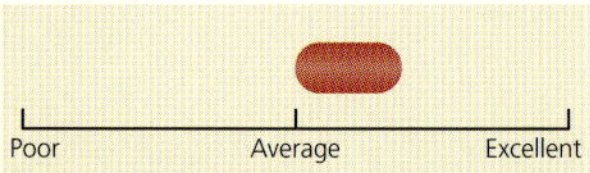

Wet soil

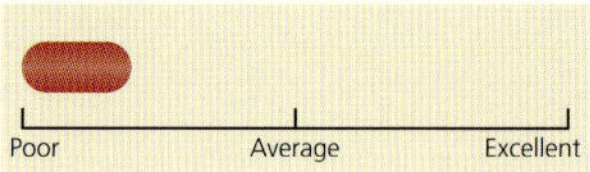

Wear

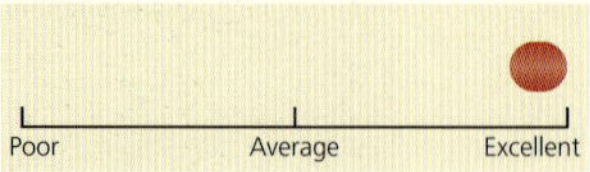

Salinity

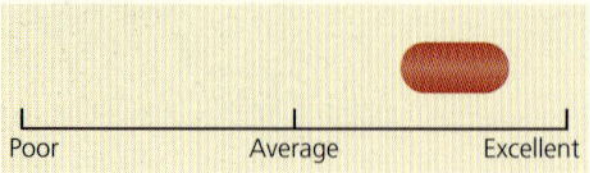

×8

×0.9

BOTANICAL NAME ***Zoysia matrella***

COMMON NAME Manila grass

OTHER NAMES Matrella zoysia

DESCRIPTION Fine textured rhizomatous and stoloniferous perennial; Leaves are tough and stiff and will form a strong turf of high durability; Leaf colour green to dark green.

COMMON USES Golf greens and surrounds, golf tees, lawns, particularly those in low light.

OTHER USES Fine turf areas likely to receive moderate maintenance.

POSITIVES Forms a very tight turf that is quite resistant to weed invasion and which can produce high quality fine surfaces; Generally good shade tolerance; Once established has high wear tolerance; Low growth rate implies lower maintenance requirement.

NEGATIVES Manila grass is slow to establish and requires extended time period for growth (although it is difficult to damage); Can form excessive thatch; Less winter hardy than Japanese lawngrass.

COMMENTS This grass is widely used for fine turf in many parts of the world and if well maintained produces a good to excellent surface; Its quality may drop if the intensity of management falls and the turf becomes excessively thatchy; Does not mix well with other grasses.

IDENTIFYING CHARACTERS Auricles – absent; Blades – stiff, hairy upper surface, smooth lower surface, acute tip, margins glabrous, width usually 2.0 to 3.0 mm.; Collar – broad, continuous, mostly glabrous but margins hairy; Ligule – fringe of hairs; Sheath – slightly compressed, glabrous, hairs at junction with leaf, split with overlapping margins; Vernation – rolled.

TOLERANCE

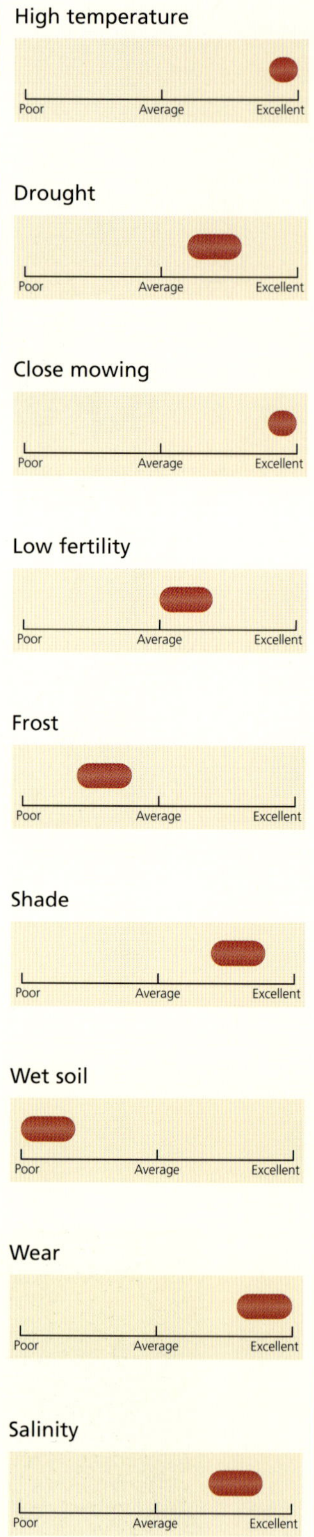

SOWING, GROWING AND MOWING

	New sowing rate		Oversowing rate		Stolonising rate		Mowing Heights	
	g/m²	lbs/1000 sq ft	g/m²	lbs/1000 sq ft	kg/100 m²	lbs/1000 sq ft	mm	inches
High	15	3.0	0	0	70	143	15	0.6
Low	10	2.0	0	0	30	61	4	0.2

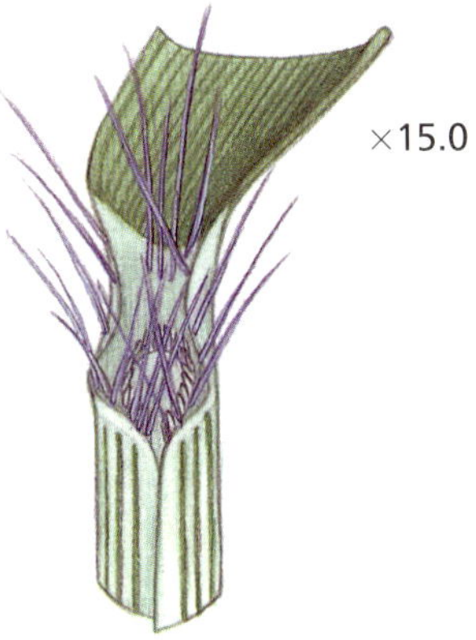
×15.0

×1.0

BOTANICAL NAME ***Zoysia tenuifolia***

COMMON NAME Korean lawngrass

OTHER NAMES Korean velvetgrass

DESCRIPTION Very fine textured perennial grass forming fine turf surface; Growth is primarily through rhizomes; Very fine leaves form a complete turf that is tough and strong; Colour green to dark green.

COMMON USES Golf greens, golf surrounds, golf tees, lawns, shaded lawns.

OTHER USES Other fine turf areas of high maintenance.

POSITIVES This grass is capable of producing high quality turf under very low mowing heights; It has some shade adaptation; Very strong once fully established; Good heat and moderate drought tolerance.

NEGATIVES Very slow rate of lateral growth means that establishment is usually by planting out sodded material rather than from stolons or seed; Very slow to repair after damage; Will form excessive thatch unless well maintained.

COMMENTS Well-known grass for use as fine turf in many tropical and subtropical zones; Can produce very high quality turf but requires significant management expertise to do so; Mixes poorly with other grasses.

IDENTIFYING CHARACTERS Auricles – absent; Blades – stiff, margins glabrous, smooth lower surface, hairy upper surface especially near the base, narrow (usually 0.5–1.5 mm); Collar – broad, continuous, mostly glabrous but margins hairy; Ligule – fringe of hairs; Sheath – slightly compressed, glabrous, hairs at junction with leaf, split with overlapping margins; Vernation – rolled.

SOWING, GROWING AND MOWING

	New sowing rate		Oversowing rate		Stolonising rate		Mowing Heights	
	g/m²	lbs/1000 sq ft	g/m²	lbs/1000 sq ft	kg/100 m²	lbs/1000 sq ft	mm	inches
High	15	3.0	0	0	70	143	10	0.4
Low	10	2.0	0	0	30	61	2.5	0.1

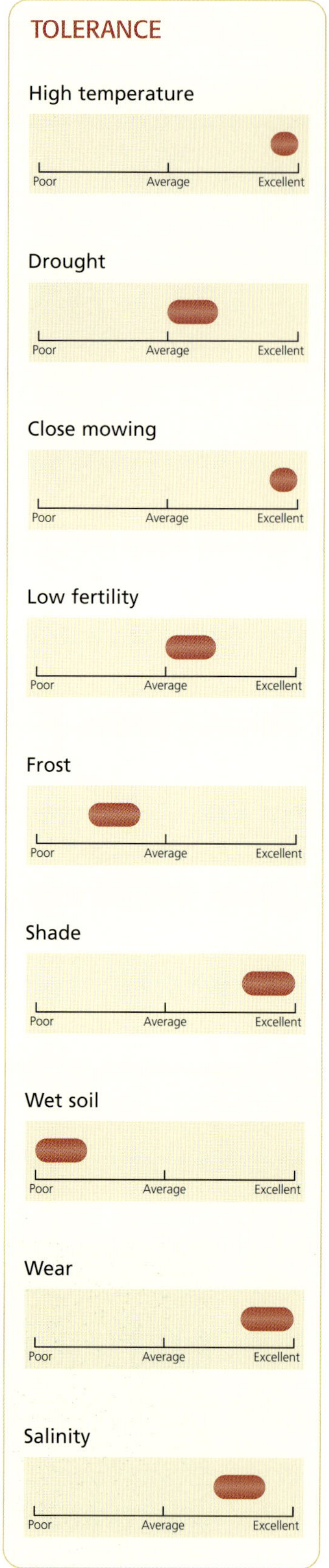

×1.0

×40.0

Minor
SPORTS TURF AND AMENITY GRASSES

The following grasses are those that are not commonly sown or planted but which regularly appear as volunteer grasses in mown turf and may comprise a significant percentage of the total turf cover.

BOTANICAL NAME *Agrostis alba*

COMMON NAME Redtop bentgrass

OTHER NAMES Black bent

DESCRIPTION Tufted perennial, with white, often long, scaly rhizomes. Spreads at the base, with erect stems. Propagated and dispersed by seed and vegetatively by rhizomes.

ADAPTATION AND COMMON USE Usually tolerant of wet, poorly drained soils, and can grow under a wide range of soils and climates. Will grow in acidic soils and is moderately salt tolerant. Becomes stemmy, coarse textured, and will soon die out under close, frequent mowing. Good seedling vigour, but not as tolerant of close mowing and has coarse leaves compared with other Bentgrasses. Turf of low shoot density, relatively coarse textured and stemmy. It is considered an undesirable species as it forms an open turf, and having strong rhizomes can be difficult to eradicate. Not recommended for use on a highly maintained turf, but performs well on roadsides, ditch and pond banks and in erosion prone areas.

IDENTIFICATION Auricles – absent; Blade – flat, stiff 3–5 mm wide, vein prominently ridged on upper surface, margins scabrous and hyaline; Collar – conspicuous, continuous, glabrous, may be oblique; Ligule – white, long, membranous, notched to acute (1.5 –4.5 mm long); Sheath – round, glabrous, split with overlapping margins, green; Vernation – rolled in bud shoot.

BOTANICAL NAME *Agrostis canina* ssp. *canina*

COMMON NAME Velvet bentgrass

OTHER NAMES Brown bentgrass

DESCRIPTION Fine textured, cool-season, stoloniferous perennial, with narrow-leafed shoots at the base, capable of forming a compact mat. Propagation principally by seed and stolons.

ADAPTATION AND COMMON USE Velvet bentgrass has been used occasionally on areas such as golf greens. Forms a dense, fine-leaved, soft turf but is not as heat tolerant as Creeping or Browntop bentgrass. Good low temperature tolerance. Performs well in low pH soils with low fertility, but does not thrive on poorly aerated and poorly drained sites. In contrast to other Bentgrasses, it will make satisfactory growth under partial shade as well as in full sunlight. Where adapted, Velvet bentgrass can be very aggressive under close, frequent mowing. It is used primarily for turf purposes such as bowling greens, putting greens and in seed mixtures for home lawns and parks.

IDENTIFICATION Auricles – absent; Blades – narrow (< 1 mm in width), flat, glabrous above and below, margins scabrous; Collar – medium to broad, continuous, glabrous; Ligule – white or almost colourless membrane, long (0.4–0.8 mm) tapering at the apex, margins smooth; Sheath – round, hairless, split with overlapping, hyaline margins; Vernation – rolled.

BOTANICAL NAME ***Austrodanthonia* spp.**

COMMON NAME Wallaby grass

OTHER NAMES Ringed wallaby grass, Satin top, White top.

DESCRIPTION Fine leafed, all-year green perennial tussock grasses, leafy at the base with erect stems. Widespread in south-eastern Australia. Propagation principally by seed and plugs.

ADAPTATION AND COMMON USE This genus contains many species that are native to Australia. The Wallaby grasses are one of the more important of the native grass genera used for pastures in the higher rainfall and temperate areas of Australia. The Wallaby grasses are cold tolerant, persistent under drought, respond well to nutrients and can withstand heavy grazing. Although highly favoured as a pasture grass, Wallaby grass produces good fairway turf on light dry soils. Very slow to establish from seed but once established can be encouraged by close mowing to produce acceptable turf. In Australia, several Wallaby grass species are being evaluated as amenity grasses under various dryland situations: *Austrodanthonia geniculata, A. setacea, A. richardsonii* and *A. caespitosa*. All show some potential for use as amenity turf.

IDENTIFICATION Auricles – absent; Blade – narrow, usually flat but often needle-like, smooth to hairy, rough on the lower surface and edges, pale or blue green; Collar – medium, continuous, hairy along the edges; Ligule – a ring of silky hairs, hairs up to 5 mm long; Sheath – narrow, grooved, and usually hairy; Vernation – folded.

BOTANICAL NAME ***Bromus catharticus***

COMMON NAME Prairie grass

OTHER NAMES Rescue grass

DESCRIPTION Bunch type, short-lived, coarse-textured, cool-season, winter and spring-growing grass which can produce annual, biennial and short-lived perennial types. Propagated principally by seed.

ADAPTATION AND COMMON USE Adapted to cool moist temperate regions of the world. Preference for deep, fertile, well-drained silt loams and clay loams. On very dry soils, Prairie grass can often exhibit an annual nature. Will not withstand frequent close mowing. Many of the *Bromus* spp. are common weeds in cultivation, roadsides and parkland, and are not recommended for turf. Prairie grass sometimes establishes in newly sown turf where its coarse prostrate stems can smother other desirable grasses.

IDENTIFICATION Auricles – absent; Blade – light green, flat, width 6–14 mm, smooth or scattered hairs; Collar – medium broad and continuous; Ligule – membranous, fairly long (about 4.0 to 6.0 mm long), sometimes covered with short downy hairs, coarsely serrate; Sheath – compressed, softly hairy (up to 2 mm long), distinctly keeled, veins obvious; Vernation – rolled in bud shoot, but may appear to be folded.

BOTANICAL NAME ***Cynosurus cristatus***

COMMON NAME Crested dog's tail grass

OTHER NAMES Dog's tail grass

DESCRIPTION Small, slender, tufted perennial grass. Propagation principally by seed.

ADAPTATION AND COMMON USE Use to a limited extent in turf and pasture mixtures in Great Britain and Europe, particularly in cool to cold, high-rainfall areas. The grass prefers moist and moderately heavy clay soils. In New Zealand, Crested dog's tail grass has been included in seed mixtures in the over sowing of pastures but has since been replaced by better grasses. In Australia, Crested dog's tail grass is of uncertain value in fine turf mixtures as it will not tolerate regular close mowing. Some breeding work is underway to develop improved varieties of this grass for use in turf conditions.

IDENTIFICATION Auricles – absent; Blade – flat and tapering to a point, 5–15 cm long, 1–4 mm wide, distinctly veined, shiny undersurface, hairless. Leaf blade has characteristic twist; Collar – narrow, continuous, glabrous; Ligule – 0.5–1.5 mm long; membranous, whitish, very blunt; Sheath – slightly flattened and yellow at the base when the dead sheaths are removed, hairless; Vernation – folded, but may be rolled in mature tillers.

BOTANICAL NAME *Dactylis glomerata*

COMMON NAME Orchard grass

OTHER NAMES Cocksfoot

DESCRIPTION Coarse-textured, cool-season, perennial bunchgrass. Propagation principally by seed.

ADAPTATION AND COMMON USE In most cool temperate countries, Orchard grass prefers moist, loamy soils, but can grow on a wide range of soil types. Has been established as a pasture species on coastal, tablelands and slopes of eastern Australia, and in waste areas in North and South Island of New Zealand, where it is used primarily in mixtures with other grasses and legumes for grazing. Exhibits shade tolerance, which makes it popular for use in orchards between the trees. Works well in erosion control mixtures. Considered a troublesome weed in turf, rather than having actual turf or amenity grass value.

IDENTIFICATION Auricles – absent; Blade – v-shaped in cross-section at base, sharply keeled below, gradually tapering to acute point, 10–45 cm long, 2–14 mm wide, deep furrow over midrib on upper surface, margins almost smooth; Collar – broad, prominent, divided, glabrous; Ligule – membranous, truncate, often with awn-like point at apex, 2–10 mm long; Sheaths – strongly compressed, flattened, keeled, green on upper surface but white below, split part way and closed below; Vernation – folded.

BOTANICAL NAME *Ehrharta calycina*

COMMON NAME Perennial veldt grass

OTHER NAMES Veldt grass

DESCRIPTION Loosely tufted, short-lived perennial grass. Leaf blade flat, yellow-green in colour, often tinged with purple. Principal means of propagation and dispersal is seed.

ADAPTATION AND COMMON USE Distributed across temperate Australia and New Zealand. Considered a grassy weed of cultivated areas. Heavy seeder and rapid coloniser of weak turf areas. Adapted to low-light situations such as at the base of evergreen trees. Often found in park land, low maintenance and waste areas. It is a native grass of southern South Africa, the seed thought to be accidentally introduced as fodder into the South Perth Zoo in Western Australia.

IDENTIFICATION Auricles – blunt, with scattered hairs; Blade – flat, glabrous, reclining; Collar – broad, glabrous, continuous; Ligule – membranous, serrated, 2–4 mm long, whitish in colour; Sheath – compressed, hairless, lower culm dull reddish-purplish colouration, keeled; Vernation – rolled.

BOTANICAL NAME ***Elytrigia repens***

COMMON NAME Twitch grass

OTHER NAMES English couch, Quack grass

DESCRIPTION Coarse textured, cool-season, sod-forming perennial grass. Propagation is by seed and rhizomes.

ADAPTATION AND COMMON USE In Europe, sometimes considered a useful component in sand fairways and amenity areas and for soil stabilisation in cool humid and semi-arid areas. In Australia, generally considered a pest grass of cultivated and waste areas. Not recommended as a turfgrass.

IDENTIFICATION Auricles – slender, clasping the stem; Blade – flat, 2.0–5.0 mm wide, rough/hairy on upper surface, smooth below, margins harsh, leaves often a bluish or grey-green; Collar – medium-broad, divided, occasional small hairs, whitish to slightly purple, v-shaped; Ligule – membranous (0.2–1.0 mm long), obtuse, may be finely toothed, short ciliate margins; Sheath – not compressed, lower sheath slightly hairy, upper sheaths smooth, veins distinct, split; Vernation – rolled in bud.

BOTANICAL NAME ***Festuca ovina***

COMMON NAME Sheep fescue

OTHER NAMES Sheep's fescue

DESCRIPTION Fine-leafed, cool-season, small, tufted perennial grass. Principal propagation by seed.

ADAPTATION AND COMMON USE Well adapted throughout cooler parts of Europe, North America and the southern cool temperate Australian States and Territories. Prefers moist, fertile soils in humid localities. Reasonable drought tolerance. Slow to establish but will persist once established. Seldom planted for turf except as a component in poor quality lawn seed mixtures. Many varieties of Hard fescue are incorrectly called Sheep fescue.

IDENTIFICATION Auricles – absent, sometimes slightly enlarged margins of the collar; Blade – bristle-like, (0.5–2 mm wide), smooth to rough above, deeply ridged on upper surface, glabrous below, pale bluish-green, margins smooth; Collar – narrow, divided, distinct, glabrous; Ligule – membranous (0.3–0.5 mm long), rounded, sparsely hairy, truncate; Sheath – round, short hairy, split with overlapping margins; Vernation – folded.

BOTANICAL NAME ***Festuca pratensis***

COMMON NAME Meadow fescue

OTHER NAMES English bluegrass, sometimes called *Festuca elatior*

DESCRIPTION Short-lived often turfed perennial bunchgrass. Principal propagation by seed.

ADAPTATION AND COMMON USE Adapted to cool temperate Europe, North America and Australasia, Meadow fescue is considered a stronger performer than Sheep fescue, being more robust, but coarser in leaf texture. Thrives best in moist fertile soils, but can make satisfactory growth on dry, sandy, gravelly or rocky soils. Adapts well to shade. Under high maintenance, the grass is often too wiry or grainy to blend well with other turfgrasses. Used to limited extent as a durable turfgrass on poorer, sandy soils, and for erosion control. Sometimes used as a cover crop for native warm-season grass and wildflower seedings. Good for erosion control.

IDENTIFICATION Auricles – small, clawlike or blunt, often brown to yellow-green, cilia present; Blade – bright green, 3–8 mm wide, glabrous and dull above, glabrous and shiny beneath, veins prominent on upper surface; Collar – continuous, broad, distinct, glabrous, pale yellow; Ligule – membranous, short (0.2–0.5 mm long), truncate to obtuse; Sheath – not compressed, not keeled, glabrous, split with overlapping hyaline margins, coloured reddish to purple at base; Vernation – rolled.

BOTANICAL NAME ***Holcus lanatus***

COMMON NAME Meadow softgrass

OTHER NAMES Yorkshire fog grass, Common velvetgrass

DESCRIPTION Coarse-textured, greyish-green, hairy, cool-season turfed perennial grass. Principal propagation by seed.

ADAPTATION AND COMMON USE Common throughout temperate Australia and New Zealand in pastures and waste places. Prefers moist soils of moderate fertility. Used as a pioneer plant for poorly drained soils. Unpalatable to stock because of hairy leaves. In North America, Meadow softgrass grows throughout the northern States but is a particular problem in the Pacific North-West. Considered too coarse for turf purposes. Other closely allied species include Creeping yorkshire fog (*Holcus mollis*) which is a soft hairy leaved rhizomatous perennial often found on poorer, acidic soils.

IDENTIFICATION Auricles – absent; Blade – flat, soft, toothed, sometimes covered with soft hairs above and below leaf surface, margins, short hairy; Collar – continuous, hairy, narrow; Ligule – membranous, medium tall, densely hairy, soft; Sheath – compressed, greyish green and bright green with distinct pink veins, densely hairy; Vernation – rolled.

BOTANICAL NAME ***Lolium multiflorum***

COMMON NAME Italian ryegrass

OTHER NAMES Biennial ryegrass, Tetilla, New Tetilla

DESCRIPTION Annual or short-lived biennial bunchgrass. Principally propagated by seed.

ADAPTATION AND COMMON USE Found in Europe, North America, New Zealand and all Australian States and Territories except the Northern Territory. Adapted to cool, moist conditions, preferring moist, fertile soils. Widespread use on roadsides and waste areas. Seed is used in low quality lawn seed mixtures, and sometimes for over-seeding warm-season perennial turfgrass swards. Other closely allied species include Wimmera ryegrass (*Lolium rigidum*) and Darnel (*Lolium temulentum*). Both are frost-resistant annuals of cropping and pasture systems, the latter being a common weed.

IDENTIFICATION Auricles – narrow, long, and claw-like (sometimes blunt); Blades – long (5–30 cm), width (3.0–7.0 mm), tapered, and upper surface dull, prominently ridged, the lower surface smooth, glossy and slightly keeled, margins smooth; Collar – broad, distinct, continuous, glabrous, smooth, pale to yellowish-green; Ligule – membranous (0.5–2.0 mm long), shorter than Perennial ryegrass. Sheath – round, not compressed or keeled hairless, base reddish purple, inner sheath yellow-green, split overlapping, hyaline margins; Vernation – rolled in the bud.

BOTANICAL NAME *Microlaena stipoides*

COMMON NAME Weeping grass

OTHER NAMES Meadow ricegrass, New Zealand ricegrass

DESCRIPTION Weeping grass is a year-long, green, low-growing perennial with a turfed or mat-forming habit. It spreads slowly under the soil by way of short, scaly rhizomes. Seed heads are lax and reclining. Considerable variations in types of Weeping grass occur with some forms capable of forming a complete turf, whereas others are too upright for turf. Does not blend well with other plants in a turf situation. Propagation by seed and plugs.

ADAPTATION AND COMMON USE Weeping grass is widespread throughout eastern and southern Australia and New Zealand. High frost and drought tolerance. Common in shaded woodlands and open forests from sea level to 670 m. Tolerates a wide range of soil types, including those of very low pH. Medium wear tolerance. Will survive relatively long periods of dry weather. Medium salt tolerance and fairly tolerant of shade. Can produce good turf if established thickly as a fairway grass and for low maintenance areas. Weeping grass responds well to increased fertility and mowing while it is growing actively. Can form a turf for golf roughs and lawns, particularly those growing under shade.

IDENTIFICATION Auricles – absent, but collar margins may overlap to suggest short blunt auricles; Blade – flat, light or bright green, fine leaves (2–5 mm wide), often outwardly curved and drooping, spear-shape leaf blade. Leaf usually rough to touch or with soft hairs. Leaf colour is usually lime green. Each side of midrib there are two well-defined veins from which there may be distinct rows of short hairs; Collar – continuous, narrow, often with tufts of silky hairs; Ligule – very short and blunt, membranous, often almost inconspicuous, fringed with a few hairs; Sheath – round, split, often with short bristly hairs near collar; Vernation – rolled.

BOTANICAL NAME ***Phleum pratense***

COMMON NAME Timothy grass

OTHER NAMES Large-leafed timothy grass

DESCRIPTION Coarse-textured, bluish-green bunched perennial grass. Lower internodes on stems enlarge to form swollen pear-shaped base. Principally spread by seed.

ADAPTATION AND COMMON USE Winter hardy. Tolerant of low temperatures. Does not tolerate drought, high temperatures, alkaline soils, or close continuous mowing. Fairly shade tolerant. An important pasture grass in Europe that can also be used for silage and conservation. Although not adapted for turf use, it has been managed on golf course fairways in Europe. Other closely allied species include Small-leafed timothy (*Phleum bertolonii*) which is a harder wearing perennial, but neither has a rhizomatous habit of growth, nor tolerates wet heavy soils.

IDENTIFICATION Auricles – absent; Blade – flat, light-greyish green, upper surface hairless, lower surface slightly rough with distinct mid vein 4–8 mm wide, margins scabrous towards base; Collar – broad, glabrous, distinct, continuous; Ligule – membranous, medium long and white, distinctly notched on either side (1.0–4.5 mm long); Sheath – smooth, rounded, hairless, pale green, nearly white at base, hyaline margins; Vernation – rolled in the bud.

BOTANICAL NAME *Poa compressa*

COMMON NAME Canada bluegrass

OTHER NAMES Flattened meadow grass

DESCRIPTION Slightly rhizomatous perennial of medium leaf texture. Leaf colour shiny and dark green with reddish margins in early spring. Distinguishable from Kentucky bluegrass which has a round stem, whereas Canada bluegrass has a flat stem, shorter leaf and longer ligule. Propagation by seed.

ADAPTATION AND COMMON USE Golf course fairways, trees and surrounds, low quality lawns, roadsides, meadows and semi-open woodland. Adapted to open, rather poor, dry soils. Can survive moderate drought stress. Less desirable than Kentucky bluegrass. Makes a poor quality turf. Some disease problems with this grass.

IDENTIFICATION Auricles – absent; Blade – short and firm, flat, scabrous on the margins, terminating in a boat-shaped tip, glabrous, keeled lower surface; Collar – narrow, divided by midrib, glabrous; Ligule – membranous, white, entire, truncate; Sheaths – slightly compressed, glabrous, green or purplish, split to the base; Vernation – folded in the bud.

BOTANICAL NAME *Poa pratensis* × *P. arachnifera*

COMMON NAME Reveille bluegrass

OTHER NAMES –

DESCRIPTION Medium-textured perennial developed between Texas bluegrass (*Poa arachnifera*) and Kentucky bluegrass (*Poa pratensis*). Dr James Read of Texas A&M University, USA, developed the successful cross in 1990 after 12 years of research. Can be distinguished from Kentucky bluegrass by its having more hairs on the joint in the inflorescence immediately below the lemmas. Propagation by seed.

ADAPTATION AND COMMON USE Reveille is a multi-use cool season grass for semi-arid regions of the southern United States. It grows well on golf courses, home lawns, commercial and industrial parks. Has an appealing green colour, is resistant to Powdery mildew (*Erysiphe graminis*) and Fall armyworm (*Pseudaletia unipuncta*), moderate resistance to Rust (*Puccinia* spp.) and White grub (*Phyllophaga* spp.).

IDENTIFICATION Auricles – absent; Blade – boat shaped tip, v-shaped to flat leaf, glabrous upper surface, both surfaces dull, glabrous margin; Collar – broad, divided, ciliate; Ligule – membranous, truncate, entire; Sheath – not compressed, glabrous, overlapping; Vernation – folded.

BOTANICAL NAME ***Poa supina***

COMMON NAME Supina bluegrass

OTHER NAMES Prostrate meadow grass

DESCRIPTION Fine textured perennial grass adapted to cool, shaded situations. Spreads by stolons and will invade under moist, shaded conditions. Propagation by seed.

ADAPTATION AND COMMON USE Native to Europe. Origin in high alps of Austria and Germany where it is suited to cool, moist conditions and heavier soils. Suited to shaded lawns, winter sports fields, golf roughs and lawns and has possible application on indoor arenas in cooler climates. Sometimes useful for shaded areas and can take moderate traffic. Good grass for cool, moist, shaded areas. Low tolerance to high temperatures and drought. Recovers poorly after drought or heat damage. Has a naturally shallow root system. Has been studied for use as indoor athletic turf under translucent roofing and has shown promise.

IDENTIFICATION Auricles – absent; Blade – fine-textured, smooth, prominent mid-vein, yellow-green, terminating in a boat-shaped tip, slightly hairy lower surface, glabrous upper surface; Collar – broad, continuous, glabrous; Ligule – membranous, short during vegetative stage, elongates as plant matures; Sheaths – compressed, glabrous, scabrous, overlapping; Vernation – folded in the bud.

BOTANICAL NAME *Puccinellia distans*

COMMON NAME Alkali grass

OTHER NAMES Weeping Alkali grass, Reflexed meadow grass

DESCRIPTION Origin in mid-European climates. Medium-fine densely tufted salt-tolerant perennial grass with winter activity. Leaf colour is dark green to bluish green, culms spreading to prostrate in open locations, erect in a sward. Can form full turf cover if sown densely. Propagation by seed.

ADAPTATION AND COMMON USE Suited for low use sports turf or lawns that are affected by salinity or the potential to be affected by salinity. Sowing rate onto soil surface 4–10 kg/ha, higher rates for heavier textured soils. Can form turf of an adequate quality under bunch type growth and can exist where other grasses cannot persist. Will establish and persist on bare saline soils and can withstand winter flooding. Not tolerant of heavy traffic or close mowing. Can often be a sparse cover, but much of this may be from unsuitable soil conditions. Can be summer dormant. On non-saline soils can act as unthrifty annual even under irrigation.

IDENTIFICATION Auricles – absent; Blade – joining sheath gradually, not stiff, shallow grooved, margins scabrous (sparsely), greyish-whitish-green, flat, tip often pink, not veined; Collar – glabrous, narrow, continuous; Ligule – membranous, obtuse-truncate, 0.9–1.7 mm long; Sheaths – glabrous, not keeled, not compressed, bleached; Vernation – folded.

BOTANICAL NAME ***Bothriochloa macra***

COMMON NAME Red grass

OTHER NAMES Redleg grass

DESCRIPTION Fine-leafed, summer growing perennial bunchgrass with basal reddish tinge. Growth habit is largely basal and spreading with seedheads protruding well above the foliage. Propagation principally by seed, but also spreads slowly by weak rhizomes.

ADAPTATION AND COMMON USE This grass is found throughout eastern and southern Australia. Red grass is tolerant of drought and heat and has low to medium frost tolerance. While not adapted to heavy shade, performance and persistence under moderate shade is good. Adapted to altitudes from sea level to 1000 m. Flowering time is from summer to autumn. Persists on a range of soils from well drained to clayey and for pH 5.8–7.5. Used primarily for grazing and conservation purposes, but has potential for non-irrigated general-purpose turf in semi-arid regions. Cultivars are being developed that will produce a fine turf surface.

IDENTIFICATION Auricles – absent; Blade – linear, rough along margins, about 3 mm, usually flat, rough to touch, leaf colour blue-green to dull dark green, often with reddish margins. Leaf blade has hairy upper surface and glabrous lower surface; Collar – broad, continuous, hairs on edge are white with reddish tinge; Ligule – short, membranous, 1.0–2.5 mm long, sometimes hairy at the edges, serrated; Sheath – glabrous, sometimes flushed with purple, slightly compressed, overlapping; Vernation – rolled.

BOTANICAL NAME *Chloris gayana*

COMMON NAME Rhodes grass

OTHER NAMES –

DESCRIPTION Rapid growing semi-perennial turfgrass spreading by long stolons rooting at the nodes. Propagation by seed, but also spread by stolons.

ADAPTATION AND COMMON USE Rhodes grass is adaptable to tropical and subtropical summer-rainfall areas with a moderately long dry season. Rhodes grass tolerates a range of soils from sands to clays, but grows best on fertile soils of medium texture. It is moderately frost-resistant, but unsuited to areas with more than an occasional heavy frost. Persistent under intensive grazing, this grass is of high pasture value and a useful coloniser. Can be used as a rough turf.

IDENTIFICATION Auricles – absent; Blade – flat, tapering, usually hairless, except for a few hairs on the upper surface near the junction of the blade and the sheath, keeled under surface, serrated edge; Collar – broad, continuous, glabrous; Ligule – a ciliate membranous rim, fringe of hairs; Sheath – very compressed, smooth, hairless, split margins, pale green; Vernation – folded in the bud.

BOTANICAL NAME ***Chloris truncata***

COMMON NAME Windmill grass

OTHER NAMES Umbrella grass, Star grass

DESCRIPTION Small, dense, perennial warm-season bunchgrass that spreads by seed and very occasionally roots at the lower nodes. Has distinctive windmill-like flower and seed heads borne on short stems above the leaves. Flowers over summer. Propagation principally by seed.

ADAPTATION AND COMMON USE Under Australian conditions, Windmill grass makes rapid growth in early spring with foliage up to a height of 20 cm. In drier areas, it may persist as an annual, regenerating from seed in spring and summer. Windmill grass demonstrates a low frost tolerance, but has a moderate drought tolerance. Useful as a rough amenity grass.

IDENTIFICATION Auricles – absent; Blade – strongly folded so stems appear flattened, hairless, light-bluish green in colour. Leaf has scabrous margin, v-shaped, deeply keeled with glabrous upper and lower surfaces; Collar – broad, divided, glabrous; Ligule – membranous, short, truncated, serrated; Sheath – very compressed and flattened, glabrous, open, pale green; Vernation – folded in the bud.

BOTANICAL NAME *Chloris ventricosa*

COMMON NAME Tall chloris

OTHER NAMES Tall windmill grass

DESCRIPTION Medium to coarse textured summer growing perennial that remains in a bunch habit of growth. Easily established from seed in warmer months.

ADAPTATION AND COMMON USE Valuable summer growing grass, favouring shaded areas. Quite drought tolerant once established. Widespread through all Australian States and Territories, and also in Europe and North America. Can tolerate regular mowing to form adequate turf for amenity purposes, although may not last for many years under these conditions.

IDENTIFICATION Auricles – absent; Blade – tapering, v-shaped, scabrous margin, keeled; Collar – broad, continuous, glabrous; Ligule – membranous, short, truncated, serrated; Sheath – very compressed, glabrous, open; Vernation – folded.

BOTANICAL NAME *Cynodon nlemfuensis*

COMMON NAME African stargrass

OTHER NAMES Star grass

DESCRIPTION Medium textured summer growing perennial, with stout and woody stolons. Leaf colour light green. Very aggressive spreading grass with long stolons and internodes up to 200 mm apart, occasionally rhizomes. Vegetatively propagated from sprigs.

ADAPTATION AND COMMON USE Easy to propagate during growing season. Cuttings establish more vigorously than Bermuda grass and can produce turf of moderate quality under regular maintenance. More drought and heat tolerant than Bermuda grass. Prefers light textured soils of moderate fertility in subtropical locations. Poor cold weather tolerance and poor growth in shade. Open canopy can permit weed invasion. Will invade nearby turf grasses. Among the best warm-season grasses for wear and recovery when grown under tropical conditions. Recuperative rate from wear good. Not wear tolerant in winter. Common uses summer sports fields, racetracks, parks and reserves and waste areas. Commercial pasture varieties available.

IDENTIFICATION Auricles – absent; Blade – flat, tapering, minute hairs on margins, both sides glabrous, neither surface glossy, veins not prominent, minutely serrated on under surface; Collar – obtuse, divided at centre, broad at sides, hairy at margins with long hairs; Ligule – fringe of hairs, hairs long 2–4 mm; Sheath – compressed, split, not overlapping, glabrous, coloured green; Vernation – rolled.

BOTANICAL NAME ***Cynodon transvaalensis***

COMMON NAME South African bermudagrass

OTHER NAMES South African couchgrass, African bermudagrass, Royal blue grass

DESCRIPTION Prostrate, fine leaf sward-forming summer growing perennial grass. Interspecific hybridisation between Bermuda grass and South African bermudagrass has produced Hybrid bermudagrass varieties such as Tifdwarf, Tifgreen and Santa Ana. Propagation can be by seed, but is mostly by stolons, rhizomes, sprigs and sod.

ADAPTATION AND COMMON USE South African bermudagrass prefers temperate and subtropical regions. Tolerates both acidic and alkaline soil conditions and flooding but does not thrive on water logged soils. South African bermudagrass is used for special purposes such as golf and bowling greens, cricket wickets, sports fields and lawn tennis courts but is considered a weed in some cool-season growing areas.

IDENTIFICATION Auricles – absent; Blade – fine, about 1–2 mm wide with 3 primary veins, not hairy, tapering; Collar – narrow, continuous, hairy; Ligule – fringe of hairs; Sheaths – slightly compressed, split, overlapping; Vernation – folded.

BOTANICAL NAME *Eragrostis cilianensis*

COMMON NAME Spreading lovegrass

OTHER NAMES Stinkgrass

DESCRIPTION Free-seeding, rapidly growing bunchgrass. The glands along its leaf margins emit a strong odour. Can be highly invasive.

ADAPTATION AND COMMON USE Naturalised to the higher rainfall areas of eastern Australia as well as the North Island of New Zealand. Can withstand drought well. Grows best in sands or other low fertility soils, but will grow in heavier soils. Develops coarse dense tufts. Despite its unattractive odour, it is palatable to stock. Not suitable for use as a turfgrass because of its short-term nature.

IDENTIFICATION Auricles – absent; Blades – fine-leaved, 3–5 mm wide, flat, hairless, often yellowish green, dull above, glossy below; Collar – continuous, narrow, long hairs at the edges; Ligule – ring of short stiff hairs; Sheaths – square leaf sheaths are crowded at the base, glossy, having prominent veins, and are often yellow near the ground; Vernation – rolled.

BOTANICAL NAME *Panicum laxum*

COMMON NAME Lax panic grass

OTHER NAMES Panic grass, Siro Shadegro grass

DESCRIPTION Short, persistent, summer growing turfed perennial. Strongly prostrate under close mowing. It does not have rhizomes, which differentiates Lax panic grass from Torpedo grass (*Panicum repens*). Propagation by seed.

ADAPTATION AND COMMON USE Adapted to shaded situations in coastal subtropics and wet tropic climes. Grows well under both high and low light. Light is required for the germination of the seed. In 1997, *Panicum laxum* cv. Shadegro was released from the University of Queensland and CSIRO Tropical Agriculture, Brisbane, Australia, for lawn and amenity use in semitropical and tropical environments.

IDENTIFICATION Auricles – absent; Blade – flat and sharply pointed, soft and pliable, glabrous or with a few scattered hairs on the upper surface; Collar – continuous, broad, and hairy; Ligule – a fringe of hairs; Sheath – slightly compressed and hairy; Vernation – rolled in the bud.

BOTANICAL NAME ***Panicum repens***

COMMON NAME Torpedo grass

OTHER NAMES Couch panic, Creeping panic grass

DESCRIPTION Erect, wiry, creeping, perennial grass, spreading extensively but not densely, from strongly developed torpedo shaped rhizomes. Propagation by means of rhizomes, and sometimes, seed. Can be a problematic weed of turf in tropical zones.

ADAPTATION AND COMMON USE Torpedo grass favours open sunny conditions, but can stand partial shade. Primarily considered a weed of moist, coastal, sandy soils, it can stand prolonged dry periods. Torpedo grass may be found along waterways and ditch banks and along the shores of lakes, often extending into the water to form large floating mats, as well as being found in pastures and sand dunes. It has potential as a coarse turf surface in warm temperate South-East Asia.

IDENTIFICATION Auricles – absent; Blade – long, tapering, sparely hairy, underside of leaf glossy, slightly keeled, margins hairy towards the base; Collar – broad, continuous, hairy at margins; Ligule – short, membranous, minutely ciliated; Sheath – not compressed, split, not overlapping, hairy at margins, reddish at nodes, veins obvious, glabrous apart from margins; Vernation – rolled in the bud.

BOTANICAL NAME *Paspalum distichum*

COMMON NAME Mercer grass

OTHER NAMES Water couch, Knotgrass

DESCRIPTION Perennial grass, which spreads by both stolons and rhizomes, to form an extensive matted sward. Stolons are usually coarse (5–10 mm wide) and have one leaf branching off at each node (except for terminal node that has 3 in close proximity). Establishment is by vegetative means only.

ADAPTATION AND COMMON USE Native to warm temperate, subtropical and tropical regions. In Australia, found along the wet-soil areas of the coast, tablelands and slopes. Mercer grass is known to prefer the littoral, warm-season growing areas where it is found in wet areas, swamps and banks of watercourses. Prefers wet, often poorly drained, non-saline soils in frost-free areas. In turf areas, Mercer grass will crowd out the finer-leaved turf grasses. Its coarse stems and leaves create a playing surface which is coarse grained and unsuitable for bowling or golf greens. Mercer grass is a particular problem in the USA from South Carolina into Florida and west to Texas and Oklahoma, but it also grows well in States west of the Rocky Mountains.

IDENTIFICATION Auricles – absent; Blade – 3–8 mm wide, open almost flat but usually margins somewhat inrolled, keeled especially near base, tapering uniformly to tip. Usually hairless on upper surface with few fine hairs beneath and on margins and near ligule. Stem nodes often hairy; Collar – narrow, continuous, with small tufts of hair; Ligule – tall (up to 4 mm long), rounded, membranous, often torn; Sheath – compressed, smooth, hairy edges, distinctly reddish-purplish tinge; Vernation – rolled.

BOTANICAL NAME ***Paspalum nicorae***

COMMON NAME Brunswick grass

OTHER NAMES –

DESCRIPTION Erect to suberect sod-forming perennial that spreads from long, branching and deep rhizomes. Propagated easily from seed.

ADAPTATION AND COMMON USE Brunswick grass prefers tropical and subtropical areas, with moderate to high well-distributed rainfall. Once established, Brunswick grass can tolerate frost. Its deep-rooted habit helps it to withstand considerable drought. Best grown in sandy soils with pH of 6.5–7.5. Brunswick grass was selected by the Plant Materials Center in Georgia, USA, for use as a cover crop and for pastures and has potential as a turfgrass in south-eastern Queensland and other tropical to subtropical climates.

IDENTIFICATION Auricles – absent; Blade – basal, aggregated in tufts at ground level, 8–12 mm broad, mostly flat or folded, rather stiff, tapering; Collar – broad, continuous, glabrous; Ligule – fringe of hairs, short, 0.2–0.3 mm long; Sheath – slightly compressed, reddish tinge at base, glabrous, open; Vernation – rolled.

BOTANICAL NAME *Sporobolus virginicus*

COMMON NAME Marine couch

OTHER NAMES Seashore dropseed, Sand couch, Beach dropseed

DESCRIPTION A fine leaved, densely turfed, salt-tolerant perennial grass. Leaves linear, narrow and blue-green in colour with prominent adaxial ribs. Principally spread vegetatively using stolons.

ADAPTATION AND COMMON USE Diverse habitats, including coastal sand dunes and intertidal saline conditions. Principally found in saline country adjacent to coastal tidal rivers on the Queensland, New South Wales, Victorian and South Australian coast. Competes well with other plants on these saline sites. Prefers sunny locations. May be found in amenity situations, such as urban parkland, roadside, caravan parks and fairways. Produces moderate quality turf when maintained at 50 mm height. Some turf varieties are entering the market.

IDENTIFICATION Auricles – absent; Blade – linear, 2–4 mm wide, hairy at junction of blade and sheath, finely toothed margins; Collar – indistinct, hairy on edge, broad, continuous; Ligule – membranous, very short, serrated; Sheath – rounded, moderately glossy, overlapping, transparent margins; Vernation – rolled.

BOTANICAL NAME *Zoysia macrantha*

COMMON NAME Prickly couch

OTHER NAMES Spiky couchgrass

DESCRIPTION Sod forming perennial grass with prostrate creeping stolons and rhizomes, found along the eastern coast of Australia and northern New Zealand. Under normal low fertility conditions, leaves tend to be quite sharp and brittle, but this can change with adequate fertility and irrigation with clean water. Propagation by sprigs.

ADAPTATION AND COMMON USE Occurs along the seashore above the high tide line where it is subject only to spray with salt water, but is not inundated. Is a natural turf-forming grass with a strong root system and great persistence under drought and under hot temperatures. Prefers lighter textured soils. Is a moderately fine-textured grass that can form good turf of useful playing quality. Moderately frost and shade tolerant.

IDENTIFICATION Auricles – absent; Blade – smooth or occasionally with short hairs above with at least a few long hairs near base, smooth below, margins smooth to scabrous; Collar – narrow, covered with hairs, continuous; Ligule – a fringe of hairs; Sheath – overlapping, round, glabrous, green coloured; Vernation – rolled in bud.

Index

African stargrass (*Cynodon nlemfuensis*), vii, 7, 102
Agrostis alba (Redtop bentgrass), vii, 9, 10, 80
Agrostis canina ssp. *canina* (Velvet bentgrass), vii, 10, 81
Agrostis capillaris (*A. tenius*) (Browntop bentgrass), vi, 7, 10, 17, 26–27
Agrostis stolonifera (Creeping bentgrass), vi, 10, 28–29
Alkali grass (*Puccinellia distans*), vii, 18, 97
Annual Meadow grass (*Poa annua*) vi, 17, 40–41
Annual poa (*Poa annua*), vi, 17, 40–41
Auricle, 1, 2
Australian buffalograss (*Stenotaphrum secundatum*), vi, 22, 70–71
Axonopus affinis (Common carpetgrass), vi, 21, 46–47
Axonopus compressus (Carpet grass), vi, 21, 48–49

Bahia grass (*Paspalum notatum*), vi, 10, 64–65
Bentgrass, Redtop (*Agrostis alba*), vii, 9, 80
Bentgrass, Creeping (*Agrostis stolonifera*), vi, 10, 28–29
Bermuda grass (*Cynodon dactylon*), vi, 22, 52–53
Blade, 1, 2
Bluegrass, Kentucky (*Poa pratensis*), vi, 20, 42–43
Bothriochloa macra (Red grass), vii, 11, 98
Bristle-like, 2
Browntop bentgrass (*Agrostis capillaris*), vi, 7, 10, 26–27
Brunswick grass (*Paspalum nicorae*) vii, 13, 108
Buffalo grass (Australia) (*Stenotaphrum secundatum*), vi, 22, 70–71
Buffalo grass (USA) (*Buchloe dactyloides*), vi, 13, 50–51
Bromus catharticus (Prairie grass), vii, 9, 83
Buchloe dactyloides (Buffalo grass), vi, 13, 50–51
Bunch type growth, 2

Canada bluegrass (*Poa compressa*), vii, 20, 94
Carpetgrass, Common (*Axonopus affinis*), vi, 21, 46–47
Carpetgrass, (*Axonopus compressus*), vi, 21, 48–49
Centipede grass (*Eremochloa ophiuroides*), vi, 19, 60–61
Chewing's fescue (*Festuca rubra* ssp. *commutata*), vi, 17, 34–35
Chloris gayana (Rhodes grass), vii, 19, 99
Chloris truncata (Windmill grass), vii, 16, 100
Chloris ventricosa (Tall chloris), vii, 17, 101
Cilia, 2
Ciliate, 2
Cocksfoot (*Dactylis glomerata*), vii, 16, 85
Collar, 1, 2
Cool season grasses, iii, vi, vii
Couchgrass, Common (*Cynodon dactylon*), vi, 22, 52–53
Creeping bentgrass (*Agrostis stolonifera*), vi, 10, 28–29
Creeping fog (*Holcus mollis*), 90
Crested dog's tail grass (*Cynosurus cristatus*), vii, 18, 84
Crown, 1
Cynodon dactylon (Bermuda grass), vi, 22, 52–53
Cynodon dactylon × *C. transvaalensis* (Hybrid bermudagrass) vi, 22, 54–55
Cynodon nlemfuensis (African stargrass), vii, 7, 102
Cynodon transvaalensis (South African bermudagrass), vii, 23, 103
Cynosurus cristatus (Crested dog's tail grass), vii, 18, 84

Dactyloctenium australe (Durban grass), vi, 22, 56–57
Dactylis glomerata (Orchard grass), vii, 16, 85
Dallis grass (*Paspalum dilatatum*), vi, 11, 62–63
Danthonia caespitosa (Common wallaby grass), vii, 23, 82
Danthonia geniculata (Wallaby grass), vii, 23, 82
Danthonia richardsonii (Wallaby grass), vii, 23, 82
Danthonia setacea (Wallaby grass), vii, 23, 82
Digitaria didactyla (Queensland blue couch), vi, 7, 58–59
Doub, Indian (*Cynodon dactylon*), vi, 22, 52–53
Dogstail grass, Crested (*Cynosurus cristatus*), vii, 18, 84
Durban grass (*Dactylocterium australe*) vi, 22, 56–57

Ehrharta calycina (Perennial veldt grass), vii, 6, 86
Elytrigia repens (Twitch grass), vii, 5, 87
Eragrostis cilianesis (Spreading lovegrass), vii, 14, 104
Eremochloa ophiuroides (Centipede grass), vi, 19, 60–61

Fescue, Chewing's (*Festuca rubra* ssp. *commutata*), vi, 17, 34–35
Fescue, Meadow (*Festuca pratensis*), vii, 6, 89
Fescue, Tall (*Festuca arundinacea*), vi, 5, 30–31
Festuca arundinacea (Tall fescue), vi, 5, 30–31
Festuca longifolia (Hard fescue), vi, 16, 32–33
Festuca ovina (Sheep fescue), vii, 16, 88
Festuca pratensis (*F. elatior*) (Meadow grass), vii, 6, 89
Festuca rubra ssp. *commutata* (Chewing's fescue), vi, 17, 34–35
Festuca rubra ssp. *rubra* (Red fescue), vi, 19, 36–37
Fog, Yorkshire (*Holcus lanatus*), vii, 12, 90

Glabrous 2
Growth habit 2

Hard Fescue (*Festuca longifolia*) vi, 16, 32–33
Holcus lanatus (Meadow softgrass), vii, 11, 90

Inflorescence, 5
Italian ryegrass (*Lolium multiflorum*), vii, 6, 91

Japanese lawngrass (*Zoysia japonica*), vi, 14, 72–73

Keel, 1, 2
Korean lawngrass (*Zoysia tenuifolia*), vi, 13, 76–77
Kentucky bluegrass (*Poa pratensis*), vi, 20, 42–43
Key, 1, 22
Kikuyu grass (*Pennisetum clandestinum*), vi, 23, 68–69

Lax panic grass (*Panicum laxum*) vii, 14 105
Leaf blade, 1, 2
Leaf margin, 1, 2
Leaf shape, 2
Leaf surface, 2
Ligule, 1, 2
Ligule margin, 2
Ligule shape, 2
Lolium multiflorum (Italian ryegrass), vii, 6, 91
Lolium perenne (Perennial ryegrass), vi, 18, 38–39
Lolium rigidum (Wimmera ryegrass), 91

Manila grass (*Zoysia matrella*), vi, 13, 74–75
Margin, 2
Membranous, 2
Midrib, 2
Meadow fescue (*Festuca pratensis*), vii, 6, 89
Meadow grass, Rough stemmed(*Poa trivialis*), vi, 20, 44–45
Mercer grass (*Paspalum distichum*), vii, 12, 107
Microlaena stipoides (Weeping grass), vii, 12, 92

Narrow-leaf carpetgrass (*Axonopus affinis*), vi, 21, 46–47
Node, 13

Orchard grass (*Dactylis glomerata*), vii, 16, 85

Panicum laxum (Siro Shadegrass), vii, 14, 105
Panicum repens (Torpedo grass), vii, 11, 106
Paspalum dilatatum (Dallis grass, Paspalum), vi, 11, 62–63
Paspalum distichum (*P. paspaloides*) (Mercer grass, Water couch), vii, 12, 107
Paspalum nicorae (Brunswick grass) vii, 13, 108
Paspalum notatum (Bahia grass), vi, 10, 64–65
Paspalum vaginatum (Seashore paspalum, Saltwater couch,), vi, 12, 66–67
Pennisetum clandestinum (Kikuyu grass), vi, 23, 68–69
Perennial, 3
Perennial veldt grass (*Ehrharta calycina*), vii, 6, 86
Perennial ryegrass (*Lolium perenne*), vi, 18, 38–39
Phleum pratense (*P. bertolonii*) (Timothy grass), vii, 9, 93
Poa annua (Annual bluegrass, Winter grass), vi, 17, 40–41
Poa compressa (Canada bluegrass), vii, 20, 94
Poa pratensis (Kentucky bluegrass), vi, 20, 42–43
Poa trivialis (Rough stemmed bluegrass), vi, 20, 44–45
Poa supina (Supina bluegrass), vii, 19, 96
Prairie grass (*Bromus catharticus*), vii, 9, 83
Pubescence, 3
Puccinellia distans (Weeping alkali grass), vii, 18, 97

Queensland Blue Couch (*Digitarie didactyla*) vi, 7, 58–59

Red fescue (*Festuca rubra* ssp. *rubra*), vi, 19, 36–37
Red grass (*Bothriochloa macra*), vii, 11, 98
Red-leg grass (*Bothriochloa macra*), vii, 11, 98
Redtop bentgrass (*Agrostis alba*), vii, 9, 10, 80
Reveille bluegrass (*Poa pratensis* × *P. arachnifera*), vii, 20, 95
Rhizome, 3
Rhode's grass (*Chloris gayana*), vii, 19, 99
Ribs, 3
Rough stalked meadowgrass (*Poa trivialis*), vi, 20, 44–45
Ryegrasses,
  Italian (*Lolium multiflorum*), vii, 6, 91
  Perennial (*Lolium perenne*), vi, 18, 38–39

Scabrous, 3
Seashore dropseed (*Sporobolus virginicus*) vii
Seashore paspalum (*Paspalum vaginatum*) vi, 12, 66–67
Serangoon grass (*Digitaria didactyla*) vi, 7, 58–59
Sheath, 1, 3
Sheep Fescue (Festuca ovina) vii, 16, 88
South African bermudagrass (*Cynodon transvaalensis*), vii, 23, 103
Sporobolus virginicus (Marine couchgrass) vii, 11, 109
St Augustine grass (*Stenotaphrum secundatum*), vi, 22, 70–71
Stenotaphrum secundatum (St Augustine grass, Buffalo grass), vi, 22, 70–71
Stinkgrass (*Eragrostis cilianensis*), vii, 14, 104
Stolons, 3
Supina bluegrass (P*oa supina*), vii, 19, 96
Sweet smother grass (*Dactyloctenium australe*), vi, 22, 56–57

Tall fescue (*Festuca arunidinacea*), vi, 5, 30–31
Tall chloris (*Chloris ventricosa*), vii, 17, 101
Tiller, 1, 3
Timothy grass (*Phleum pratense*), vii, 9, 93
Torpedo grass (*Panicum repens*), vii, 11, 106
Twitch grass (*Elytrigia repens*), vii, 5, 87

Velvet bentgrass (*Agrostis canina* ssp. *canina*), vii, 10, 81
Venation, 3
Vernation
  folded, 3
  rolled, 3

Yorkshire fog (*Holcus lanatus*), vii, 12, 90

Wallaby grasses (*Austrodanthonia* spp.), vii, 23, 82
Weeping alkali grass (*Puccinellia distans*), vii, 18, 97
Weeping grass (*Microlaena stipoides*), vii, 12, 92
Windmill grass (*Chloris truncata*), vii, 16, 100
Winter grass (*Poa annua*), vi, 17, 40–41

Zoysia japonica (Japanese lawngrass), vi, 14, 72–73
Zoysia macrantha (Prickly couch), vii, 14, 110
Zoysia matrella (Manila grass), vi, 13, 74–75
Zoysia tenufolia (Korean lawngrass), vi, 13, 76–77